# INJECTORS:

THEIR

# THEORY, CONSTRUCTION, AND WORKING

BY

W. W. F. PULLEN,

*Wh.Sc., Assoc.M.Inst.C.E., M.Inst.M.E.; Assistant Lecturer and Demonstrator in Engineering at the University College of South Wales and Monmouthshire, Cardiff.*

---

UPWARDS OF 100 ILLUSTRATIONS.

---

1893.
THE TECHNICAL PUBLISHING COMPANY LIMITED,
6, VICTORIA STATION APPROACH, MANCHESTER.
JOHN HEYWOOD,
DEANSGATE AND RIDGEFIELD, MANCHESTER.
2, AMEN CORNER, LONDON, E.C.
33, BRIDGE STREET, BRISTOL.
*And all Booksellers.*

*

## APROPOS ............

The workings of the steam injector, and its sister, the steam ejector are, to many, the great mystery of steam power.

This book may be over 100 years old, but its coverage of the subject, and in particular of design parameters, is better than any other we have seen. We believe it will go a long way towards making the workings of injectors and ejectors clearer, and help those who wish to make their own injectors, or repair and maintain them.

Judging by the names in the original copy from which this book is reprinted, this had at least three owners, one or more of whom has added their own footnotes, and these have been left as they were added, notably as an extra page between original pages 40 and 41.

---

Published in Great Britain by:
Camden Miniature Steam Services,
Barrow Farm, Rode,
Nr. Frome, Somerset. BA11 6PS
*www.camdenmin.co.uk*

Originally published 1893 by
The Technical Publishing Company

First reprinted 1997


ISBN 978-0-9519367-5-7



British Library Cataloguing-in Publication Data.
A catalogue record for this book is available from the British Library.

# PREFACE.

In this work it has been thought advisable to stray somewhat from the well-trodden path so generally taken under similar circumstances, and to reserve the historical portion until the theory has been discussed at some length. With this arrangement, it is the Author's opinion, the general reader is placed in a more fitting position to follow and appreciate the different stages of development; and carrying out this idea, this book *concludes* with an historical summary of the Injector and cognate apparatus.

The first chapter opens with a popular explanation of the action of the steam injector. Immediately following will be found the mathematical investigation of the velocity of efflux of the steam jet; the different steps throughout this and the remainder of the work being given with much consecutiveness, more especially for the purpose of assisting those who may not be quite familiar with the mathematics there used in following the reasoning.

To permit those who are not able to follow the latter part of Chapter I., a table has been constructed at the end of Chapter II., giving the principal data and results which may be needed in any of the calculations that come after.

The method of treatment given in Chapters VI. and VII. is due to Professor Elliott, D.Sc., much of it being taken from his paper there mentioned, which he kindly placed in the Author's hands.

In the rather large number of illustrations which follow, the Author has endeavoured to give examples of the different forms rather than examples of each manufacturer's apparatus. The generosity of makers and patentees in supplying blocks for the illustrations, where possible, has rendered the preparation of this work much less arduous than it otherwise would have been, and the Author desires to express his thanks to them.

It will, no doubt, be noticed that of the examples given very few are of American origin, which fact is due to the difficulty of obtaining them, together with the impossibility of consulting any American patent records.

Although the exhaust injector is identically similar to the ordinary live-steam apparatus, except in dimensions, it has been thought desirable to treat it separately, together with compound injectors. The same applies to the ejector condenser.

In the historical portion, the dates given have been almost exclusively taken from the patent specifications, as have also many of the illustrations ; and it may here be remarked that much useful information may be obtained from the volume of abridgments of patent specifications (Injectors) recently published by the Patent Office.

The only treatise published in this country in which the Author has been able to find more than a passing reference to the injector is Professor Peabody's admirable work on Thermodynamics, which the Author read with much interest soon after its publication.

To endeavour to render any portion of this book readable at any time, the different equations have all been numbered consecutively, and copious references have been made to them throughout. Also an index of symbols has been made, so that the reader may, by referring to it, interpret any equation or formula, without having to first wade through the previous pages to find out what each symbol represents.

W. W. F. P.

University College,
Cardiff, 1893.

# INDEX.

---

ERRATA.—Page 9, line 11, for $PV_n$, read $PV^n$ = constant. Page 47, for number of injector and diameter of delivery cone in *inches*, read *millimetres*.

# INDEX OF SYMBOLS.

---

THE suffix ($a$) refers to the fluid at atmospheric pressure.
The suffix (1) refers to the fluid in the boiler.
The suffix (2) refers to the fluid in the steam orifice.
The suffix (3) refers to the fluid in delivery orifice.
The suffix (4) refers to the fluid in final delivery chamber (generally the boiler).
The suffix (5) refers to the fluid in the feed-water pipe.
The suffix (6) refers to the fluid at surface of feed-water tank.
The suffix (8) refers to the fluid in final delivery chamber of a compound injector.

$A$ = area in square feet.
$P$ = pressure in pounds per square foot.
$p$ = pressure in pounds per square inch.
$v$ = velocity in feet per second.
$V$ = volume in cubic feet.
$J$ = Joule's mechanical equivalent 772.*
$I$ = intrinsic energy = total heat − external work.
$x$ = dryness fraction—*i.e.*, that fraction of the whole weight of fluid which is dry saturated steam.
$h$ = sensible heat—*i.e.*, heat required to raise the temperature of one pound of water from 32 deg. Fah. to $t$ degree.
$L$ = latent heat of steam—*i.e.*, that amount required to convert one pound of water into steam without altering its temperature $t$.
$s$ = volume of one pound of water = ·016 cubic foot approximately.
$\rho$ = density—*i.e.*, the weight of one cubic foot of fluid.
$t$ = temperature on the Fahrenheit thermometer.
$T$ = absolute temperature (Fah. scale).
$n$ = the exponent in the equation $P V^n$ = constant.
$H$ = height of surface of cold feed water above the injector.

$$\beta = \text{ratio} - \frac{\text{weight of cold feed water}}{\text{weight of steam used}}$$

$r$ = radius of cone orifice in inches.
$d$ = diameter of cone orifice in inches.
$f$ = coefficient of fluid resistance (page 34).

$$z = \left(\frac{2}{n+1}\right)^{\frac{1}{n-1}} \sqrt{2g \cdot \frac{n}{n+1}} \quad \text{equation (42.)}$$

$$y = \frac{A_5}{A_2}$$

$$w = \frac{n P_2}{P_1} \quad \text{equation (40).}$$

$$\beta_1 = \text{ratio } \frac{\text{water in lifter}}{\text{steam in lifter}} \quad \text{page 91.}$$

$$\beta_2 = \text{ratio } \frac{\text{water supplied to forcer from lifter}}{\text{steam supplied to forcer.}}$$

---

* Recent determinations by Rowlands and others, give 780 instead of 772.

# THE INJECTOR.

## CHAPTER I.

If one happens to glance through most of the books devoted to the study of the steam engine, as a whole or in part, he must be forcibly struck with the paucity of information on the subject of "Injectors." Putting aside the curiosity with which most persons regarded the instrument on their first introduction to it, and the apparently paradoxical result obtained with it, it is indeed difficult to find sufficient reason to account for this want of consideration for an instrument so useful in purpose and so beautiful in its action.

A machine we can hardly call it, from a strictly kinematical point of view; but there are few instruments in the every-day use of the engineer the result of whose working can be calculated with such nicety as that of the injector. Moreover, it is an instrument which shows the simple application of the grand principle of the *conservation of energy.* We are too apt to look upon this universal principle as belonging more especially to astronomy and cognate subjects, and to ignore the fact that the old rule—the weight multiplied by its arm equals the power multiplied by its arm—is but a simple example of that principle applied to the work of the mechanic. The fact that *action and reaction are equal and opposite* wherever we find the phenomenon force in nature is hardly recognised to the extent that it should be, although, when we come to soberly think over the matter, we fail to see how it could be otherwise. We read in our text books that *force* is proportional to, and is measured by, the *momentum it will generate in a unit of time;* and we are too often guilty of burying the words deeply in our memories for the purpose of enabling us to write them out again in the form of a definition in an examination, rather than thinking over them, and applying them to some useful purpose. By selecting proper units we may say that force *equals* the momentum it will generate in a unit of time. Two or more bodies can mutually act upon one another, and this mutual action we call force; in fact, the only way in which one body can act upon another is by means of force. These mutual actions (forces) must of necessity be equal; and as force equals the momentum it generates, therefore the mutual momenta (if we may so call them) must be equal and opposite in direction, or, in other words, the total momentum (the sum of both) must be a quantity which is unaffected by the mutual actions of the bodies forming the same system. This is sometimes called the conservation of momentum, and is completely expressed by Newton's third law. If two or more bodies collide, then it follows directly from the above that the total momentum (sum of momenta of bodies) in the line of action remains unaltered by the impact, or, in other words, the *total momentum before impact equals the total momentum after impact.*

We have seen that momentum is generated or destroyed by force; hence there must be some relation between the force, the momentum, and the time in which the force generates the momentum. When dealing with the unit of time or the second, we have shown that

$$\text{Force} = \text{mass} \times \text{velocity generated in the mass per second.}$$

If, now, the force acts for a longer or shorter period than one second, then multiply both sides of the equation by the time during which the force acts, and we get:

$$\text{Force} \times \text{time} = \text{mass} \times \text{velocity generated in one second} \times \text{time} = \text{mass} \times \text{velocity generated in } t \text{ seconds}$$

where $t$ = the time in seconds.

This equation holds, whether the time during which the force is acting is a quarter of an hour or the millionth part of a second; hence we may make use of it equally well in the case of impact, where the time of action is very small, as in cases where the action is more prolonged.

In the above equation the left-hand side (force × time) is generally called the *impulse;* and the right-hand side (mass × velocity) the momentum generated or destroyed by the impulse. As *impulse* and *momentum* are of the same dimentions, we may, if we like, add different impulses to other momenta, or subtract one from the other, and the result will also be impulse or momentum. This we shall require to make use of in the investigation of the working of the injector.

In the few notes that follow the writer does not pretend to claim originality for any part of them, but it has occurred to him that a more lucid exposition is necessary of what has already been done, and after his remarks at the beginning he thinks no further excuse for starting this article is necessary.

Information and data are generally troublesome to obtain, and manufacturers must find it no small nuisance to be continually bothered for information regarding their several manufactures. The writer has been favoured with courteous replies to his inquiries for details, &c., for this article, and at the outset he is desirous to express his grateful acknowledgment and thanks to those who have so kindly assisted him by supplying sketches and details.

Before entering upon the mathematical analysis, we will first see if any popular explanation of the apparent paradoxical result can be found, such as will satisfy the inquiries of the lad who, by accident or force of circumstances, soon

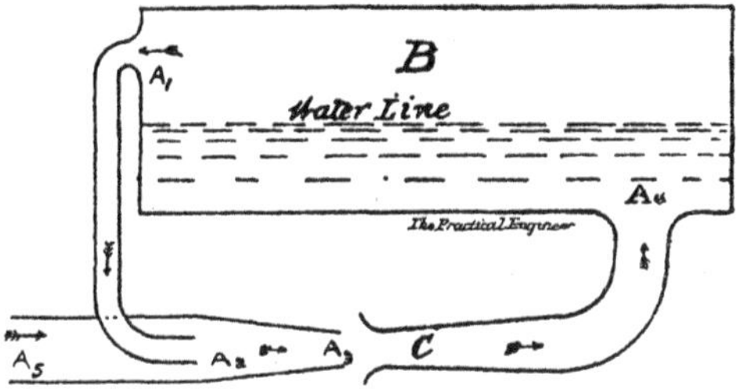


Fig. 1.

after entering the workshop, has had his attention directed to the instrument in question. The sketch, fig. 1, represents diagrammatically a section of the injector and the boiler from which it derives its motive power, and to which it delivers the feed water.

The cylindrical vessel B is the boiler, having above the water line an aperture $A_1$, to which is connected a steam pipe $A_1$ $A_2$, terminating in an orifice at $A_2$. A pipe $A_5$ connected to a reservoir admits of cold feed water passing from the reservoir to the orifice $A_2$ of the steam pipe; while a third pipe C, having an orifice near $A_3$, terminates at $A_4$ on entering the boiler. For the present all valves or stop taps are left out of the question. We have at the orifice $A_1$ steam at a pressure of say $p_1$ pounds per square inch, the same as that of the boiler. We also have at the orifice $A_4$ a pressure of $p_1$ pounds on the square inch. Now, the steam flows down the pipe $A_1$ $A_2$ as a stream of steam, having a certain area of cross section, say $a_2$, until it comes in contact with the cold water at the orifice $A_2$. Here the steam will be condensed—that is, the cross section will become much less than before ; but still the velocity remains the same as that in the pipe $A_1$ $A_2$. Let the area of the cross section of the stream after condensation be $a_3$. The stream of steam between $A_1$ and $A_3$ resembles a tapering rod whose cross section at $A_1$ is $a_1$, and at $A_3$ is $a_3$ (which is very much less than $a_1$). If now we have a pressure of $p_1$ pounds per square inch on the end $A_1$, the total pressure on $A_1$ is $p_1 \times a_1$ pounds. Let the equivalent pressure on the other end of the rod be $p_3$ pounds per square inch, then the total pressure on that end is $p_3$ $a_3$ pounds. Now, as the *same weight of matter* passes down *any* portion of the steam pipe in the same time, and as the velocity of that matter is unaltered by the condensation of the steam, the total pressure on one end of our hypothetical rod must be the same as the total pressure on the other end ; hence $p_1 a_1 = p_3 a_3$, and therefore

$$p_1 \frac{a_1}{a_3} = p_3$$

Now, $a_1$ is much greater than $a_3$ ; therefore $p_3$ is much greater than $p_1$. The condensed steam (neglecting the feed water for the moment) passes with the same velocity, and therefore with the same equivalent pressure, into the orifice of the tube C, where we have the following phenomenon—an equivalent pressure of $p_3$ pounds per square inch at one end and a pressure of $p_1$ pounds per square inch at the other end. Above it has been shown that $p_3$ is much greater than $p_1$ ; hence, as the greater intensity of pressure must prevail, the stream of condensed steam will drive back the stream of water coming out of the boiler at $A_4$. If an experiment were made, it would be found that the equivalent pressure $p_3$ would be many times $p_1$, and therefore there would be a large margin of pressure available for the purpose of carrying with the condensed steam into the boiler the cold water which condensed it.

The writer has thought it perfectly legitimate to use the term *equivalent pressure* when applied to a moving jet of fluid. The novice can easily satisfy himself that a pressure can be produced by a jet of water, by causing a jet to play horizontally on the scale pan of an ordinary spring balance, the line of direction of the jet being perpendicular to the scale pan. The pressure of the jet will then be registered on the dial.

The above explanation of the action of an injector perhaps does not represent *exactly* everything that takes place, but it is enough to show anyone who is not well versed in mechanical science that the so-called paradox is only apparent, and that there is really no difficulty in conceiving a boiler to be thus fed, by steam taken from itself and made to pass through proper fixed channels, without any additional moving parts by which a mechanical advantage is usually obtained.

To get definite numerical results, it is necessary to determine the velocity with which steam issues from the orifice of a vessel, where it is contained under a given pressure ; and this we now propose to do in two ways, by two slightly different applications of the same principle, namely, that of the "conservation of energy."

In figure 2, let the vessel X terminate at one end in a comparatively small pipe Y, and, for the sake of simplicity, let both pipe and vessel be cylindrical. Also let the vessel

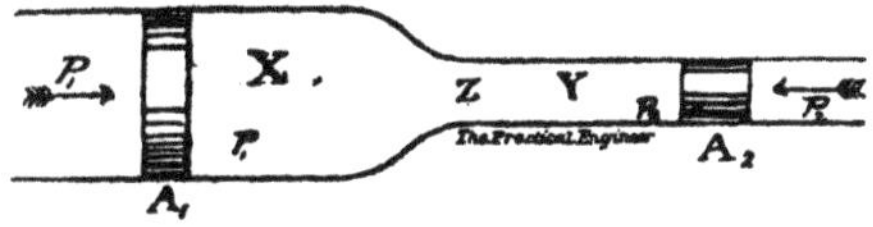


FIG. 2.

X be fitted with a fluid-tight piston without fricton, and a similar piston in the pipe Y. The pipe and vessel are supposed to be filled throughout with fluid, which moves, say, from left to right, there being a constant pressure $P_1$ pounds per *square foot* in X, and a constant pressure $P_2$ pounds per square foot in Y, on both sides of the pistons ; so that the pistons themselves are identical with the quantities of fluid they displace. Also let the motion be "steady"—that is, the fluid moves in stream lines, and the quantity which passes through every section of the pipe or vessel in any particular time must be the same, or constant, for that time and the given pressures. This last condition is sometimes called the "condition of continuity," and may be expressed by an equation thus: Let A represent the area of cross section of the pipe or vessel, and $v$ the linear velocity of the fluid at the cross section A. Also let $\rho$ be the density of the fluid at the same section—*i.e.*, the weight of the unit of volume—and the suffixes 1, 2, &c., indicate the several quantities at the particular sections 1, 2, &c. Then,

since the weight that passes any and every section is the same in a given time, we have—
Weight flowing per second

$$= A\,v\,\rho = A_1\,v_1\,\rho_1 = A_2\,v_2\,\rho_2 = \text{constant} \quad . \quad . \quad . \quad (1)$$

By the principle of the conservation of energy, the sum of the kinetic and potential energies of a body, or system of bodies, is a constant quantity. Hence, if the potential energy decreases, the kinetic energy must increase through the same amount; and *vice versa*. This is sometimes given as: The work done upon a body equals the gain in kinetic energy. Let V represent the volume of 1lb. weight of fluid in cubic feet. Then, when 1lb. of fluid passes from X into Y, we have: Work done *upon* the pound of fluid in X = $P_1\,V_1$; work done *by* the pound of fluid in Y in forcing along the small piston against a pressure $P_2 = P_2\,V_2$. Hence net work done upon the pound of fluid, or the gain of energy by the pound of fluid = $P_1\,V_1 - P_2\,V_2$.

Let the total internal or intrinsic energy of the pound of fluid in X be represented by $I_1$, and the intrinsic energy of the same pound of fluid when it has passed into Y be $I_2$. It is convenient that these last quantities should be expressed in thermal units; hence, to bring them to units of work, we must multiply them by J (Joule's equivalent, 772 foot-pounds per British thermal unit). Then the intrinsic energy converted into work, or change of intrinsic energy of the pound of fluid in passing from X to Y, = $J\,(I_1 - I_2)$.

Again, if $v_1$ is the velocity of fluid in X, and $v_2$ that in Y, the change, or gain of kinetic energy of the pound of fluid in passing from X to Y, is—

$$\frac{1}{g}\,v^2{}_2 - \frac{1}{g}\,v_1{}^2;$$

or,

$$\frac{v_2{}^2 - v_1{}^2}{2\,g} \text{ foot-pounds.}$$

Also let Q units of heat pass into each pound of fluid as it passes from X to Y.

Now, as the change in the potential energy must be numerically equal to the change in the kinetic energy (from the principle of the conservation of energy), we find, after collecting the several quantities above, that—

$$P_1\,V_1 - P_2\,V_2 + J\,(I_1 - I_2) + J\,Q = \frac{v_2{}^2 - v_1{}^2}{2\,g} \quad . \quad . \quad (2)$$

If the fluid be water at ordinary temperatures, then $I_1 = I_2$, and $Q = O$; also the weight of a cubic foot = 62·5lb.; and therefore the volume of 1lb.

$$= \frac{1}{62·5} \text{ cubic feet} = \frac{1}{\rho}.$$

Then, as $V_1 = V_2$, the above equation becomes—

$$\frac{P_2}{\rho} + \frac{v_2^2}{2g} = \frac{P_1}{\rho} + \frac{v_1^2}{2g} = \text{constant} \quad . \quad . \quad . \quad (3)$$

If the fluid be ordinary steam, let $x$ denote that fraction of each pound of fluid which is real saturated steam, and then $1 - x$ must be the water held in suspension as water by the steam. The quantity $x$ is called the dryness fraction. Also let $h$ denote the thermal units required to raise 1lb. of water from 32 deg. Fah to $t$ deg. Fah.; and let L be the number of thermal units required to convert 1lb of. water, at a temperature of $t$ deg. Fah. into saturated steam at $t$ deg. Fah.; in other words, L is the latent heat. Let $s$ = the volume of 1lb. of water. This is approximately constant, and equals ·016 cubic feet. The intrinsic energy of steam (reckoned from 32 deg.) = total heat of evaporation − external work done during the formation. Therefore—

$$J\,I_1 = J\,(h_1 + x_1 L_1) - P_1 (V_1 - s) \quad . \quad . \quad . \quad (4)$$

$$J\,I_2 = J\,(h_2 + x_2 L_2) - P_2 (V_2 - s) \quad . \quad . \quad . \quad (5)$$

Let $V^1$ be the volume of 1lb. of *saturated steam*, corresponding to a pressure of P pounds per square foot. Then, as volume V contains part steam and part water, we have—

$$V = x V^1 + (1 - x)\,s = x\,(V^1 - s) + s \quad . \quad . \quad (6)$$

Therefore, $$(V - s) = x\,(V^1 - s).$$

Putting this value in equations (4) and (5), we obtain—

$$J\,I_1 = J\,(h_1 + x_1 L_1) - P_1 x_1 (V^1_1 - s)$$

$$J\,I_2 = J\,(h_2 + x_2 L_2) - P_2 x_2 (V^1_2 - s)$$

Also substituting from equation (6), we find that—

$$P_1 V_1 - P_2 V_2 = P_1 [x_1 (V_1^1 - s) + s] - P_2 [x_2 (V^1_2 - s) + s] \quad . \quad (7)$$

Returning now to equation (2), and replacing the several quantities by their equivalent in equations (4) to (7), the result we obtain is that—

$$\frac{v_2^2 - v_1^2}{2g} = (P_1 - P_2)\,s + J\,(h_1 - h_2 + x_1 L_1 - x_2 L_2) + J\,Q \quad (8)$$

In the following pages, $v_1$, the initial velocity, will be very small indeed compared with $v_2$, and hence it can be neglected; also the flow of fluid is supposed adiabatic—that is, without any heat passing into or out of the fluid during motion; hence $Q = O$.

Now, if $x_1$ is known, we may find $x_2$ from the equation of adiabatic expansion of steam

$$\frac{x_1 L_1}{T_1} + \phi_1 = \frac{x_2 L_2}{T_2} + \phi_2$$

where T is the absolute temperature and $\phi$ the entropy. If tables of entropy are not at hand, the following approximate equation may be used, in which the specific heat of water is supposed constant:

$$\frac{x_1 L_1}{T_1} + \log_e T_1 = \frac{x_2 L_2}{T_2} + \log_e T_2 \quad . \quad . \quad . \quad (9)$$

from which we see that

$$x_2 = \left[\frac{x_1 L_1}{T_1} + \log_e \frac{T_1}{T_2}\right]\frac{T_2}{L_2}$$

Equation (8) now becomes—

$$\frac{v_2^2}{2g} = (P_1 - P_2)\,s + J\left[h_1 - h_2 + x_1 L_1\left(1 - \frac{T_2}{T_1}\right) - T_2 \log_e \frac{T_1}{T_2}\right] \quad . \quad . \quad . \quad . \quad . \quad . \quad (10)$$

This equation is applicable to all possible mixtures of steam and water from dry saturated steam down to pure hot water. A series of experiments, carried out by Napier before 1869, goes to show that when steam is flowing from an orifice, the pressure in that orifice is never less than that which would produce the maximum flow of weight of steam from the orifice; and Rankine, after an examination of the results of the experiments, came to the same conclusion. Hence, for our calculations relating to injectors we shall require to find the ratio of the pressure in the steam orifice to that in the boiler, so as to produce the maximum weight flow of steam.

It will be shown a little further on that this ratio $\frac{P_2}{P_1}$ is about ·6. The above results have been deduced from a consideration of the dynamical and thermal properties of the fluid taken together, but it is equally as interesting to obtain a similar result from purely dynamical reasoning, which the writer proposes to do in what immediately follows. Writing down again equation (2) in the form—

$$\frac{P_1 - P_2}{\rho} = \frac{v_2^2 - v_1^2}{2g}$$

we see from inspection that the left-hand side is the integral of $\frac{dP}{\rho}$ between the limits 1 and 2; hence the differential equation may be written without limits thus:

$$\int \frac{dP}{\rho} + \frac{v^2}{2g} = \text{constant} \quad . \quad . \quad . \quad . \quad . \quad . \quad (11)$$

This is the differential equation of steady motion for any fluid, whether liquid or gaseous. Let the pressure and volume be connected by the equation,

$$P V_n = \text{constant} = K, \text{ say,}$$

then, as $V = \frac{1}{\rho}$, we have the equation—

$$P = K \rho^n \quad . \quad . \quad . \quad . \quad . \quad . \quad . \quad . \quad (12)$$

Differentiating, we have—

$$dP = n K \rho^{n-1} d\rho,$$

and

$$\frac{dP}{\rho} = \frac{n K \rho^{n-1} d\rho}{\rho} = n K \rho^{n-2} d\rho.$$

After integrating, we obtain—

$$\int \frac{dP}{\rho} = \int n K \rho^{n-2} d\rho = \frac{n}{n-1} K \rho^{n-1}.$$

Multiply numerator and denominator of right-hand side by $\rho$; then, remembering that $P = K \rho^n$, we obtain

$$\int \frac{dP}{\rho} = \frac{n}{n-1} K \frac{\rho^n}{\rho} = \frac{n}{n-1} \frac{P}{\rho} \quad . \quad . \quad . \quad . \quad (13)$$

Putting in the limits, and substituting the result in equation (11), we find that

$$\frac{n}{n-1}\left[\frac{P_1}{\rho_1} - \frac{P_2}{\rho_2}\right] = \frac{v_2^2 - v_1^2}{2g} \quad . \quad . \quad . \quad . \quad (14)$$

Let the initial velocity $v_1$ be zero and put in the volumes for the inverse of the densities, then (14) becomes

$$\frac{n}{n-1}\left[P_1 V_1 - P_2 V_2\right] = \frac{v_2^2}{2g}$$

The left-hand side of this equation represents the actual work done by 1lb. of fluid in expanding adiabatically from $P_1$ to $P_2$. We may therefore naturally conclude that all the work spent is used in producing velocity in the pound of the fluid. The right-hand side is the kinetic energy of 1lb. of fluid moving with a velocity $v_2$ feet per second. This last equation is merely a return to the principle of the conservation of energy, or the work spent = kinetic energy generated.

Let Q = the weight flow per second

$$= A\,v\,\rho = A_1\,v_1\,\rho_1 = A_2\,v_2\,\rho_2 \text{ from equation (1),}$$

substituting the value $v_2$ from equation (14), remembering $v_1 = 0$.

$$Q = A_2\,\rho_2\sqrt{2\,g\,\frac{n}{n-1}\left(\frac{P_1}{\rho_1} - \frac{P_2}{\rho_2}\right)} \quad . \quad . \quad . \quad . \quad (15)$$

This expression will be a maximum for a given value of

$$\frac{P_1}{\rho_1} \text{ when } \frac{P_1}{\rho_1}\rho_2^2 - P_2\,\rho_2$$

is a maximum. For $P_2$ put $K\rho^n$, then the above expression becomes, after differentiating and equating to zero,

$$\frac{2P_1}{\rho_1}\rho_2 - (n+1)\,\rho_2^n = 0$$

$$\text{or } \frac{P}{\rho_1} = \left(\frac{n+1}{2}\right)\frac{P_2}{\rho_2} \quad . \quad . \quad . \quad . \quad . \quad . \quad . \quad . \quad (16)$$

Putting this value of $\frac{P_2}{\rho_2}$ in equation (15), we get

$$Q \text{ max.} = A_2\,\rho_2\sqrt{2g\frac{n}{n-1}\left[\frac{P_1}{\rho_1} - \frac{2}{n+1}\cdot\frac{P_1}{\rho_1}\right]}$$

$$= A_2\,\rho_2\sqrt{2\,g\frac{n}{n+1}\frac{P}{\rho_1}}$$

$$= A_2\,\frac{\rho_2}{\rho_1}\sqrt{2\,.\,g\,.\frac{n}{n+1}\,.\,P_1\,\rho_1}\,. \quad . \quad . \quad (17)$$

From equation (16) we have—

$$\frac{\rho_2}{\rho_1} = \left(\frac{n+1}{2}\right)\frac{P_2}{P_1} = \frac{n+1}{2}\left(\frac{\rho_2}{\rho_1}\right)^n$$

Divide through by $\frac{\rho_2}{\rho_1}$, and we get

$$\frac{2}{n+1} = \left(\frac{\rho_2}{\rho_1}\right)^{n-1}$$

$$\text{or} \left(\frac{2}{n+1}\right)^{\frac{1}{n-1}} = \frac{\rho_2}{\rho_1} \quad . \quad . \quad . \quad . \quad . \quad . \quad (18)$$

Substituting this value in equation (17),

$$Q \text{ max.} = A_2 . \left(\frac{2}{n+1}\right)^{\frac{1}{n-1}} \sqrt{2g \frac{n}{n+1} . P_1 \rho_1} \quad . \quad . \quad (19)$$

This last expression for the maximum weight flow of steam corresponds with what we find by experiment, and it is now only left to determine the ratio of the pressures which produce this maximum flow.

Reverting to equation (16) for the relation between pressures and densities for maximum weight flow, we find that

$$\frac{\rho_2}{\rho_1} . \frac{2}{n+1} = \frac{P_2}{P_1}$$

and putting in the value of $\frac{\rho_2}{\rho_1}$ from equation (18),

$$\frac{P_2}{P_1} = \frac{2}{n+1} . \left(\frac{2}{n+1}\right)^{\frac{1}{n-1}} = \left(\frac{2}{n+1}\right)^{\frac{n}{n-1}} . \quad . \quad . \quad . \quad (20)$$

If we now put $n = 1$, the ratio is in an undetermined form, and we require to find its true value. This is easily done as follows :—

$$\text{Let } \frac{n+1}{2} = Z + 1, \text{ then } Z = \frac{n-1}{2}.$$

$$\text{also let } a = \frac{n}{1-n}$$

We may write equation 20 thus :

$$\frac{P_2}{P_1} = \left(\frac{n+1}{2}\right)^{\frac{n}{1-n}} = (1+Z)^a$$

Expanding by the binomial theorem, we get—

$$(1+Z)^a = 1 + aZ + Z^2 \frac{a(a-1)}{1 \times 2} + Z^3 . \frac{a(a-1)(a-2)}{1 \times 2 \times 3} + \&c.$$

$$= 1 + a . \frac{n-1}{2} . + \left(\frac{n-1}{2}\right)^2 . \frac{a(a-1)}{1 \times 2} + \left(\frac{n-1}{2}\right)^3 .$$

$$\frac{a(a-1)(a-2)}{1 \times 2 \times 3} + \&c.$$

$$= 1 + \frac{n}{1-n} \cdot \frac{n-1}{2} + \left(\frac{n-1}{2}\right)^2 \cdot \left[\frac{n}{1-n} - 1\right] \frac{1}{1 \times 2} +$$

$$\left(\frac{n-1}{2}\right)^3 \frac{n}{1-n} \cdot \left[\frac{n}{1-n} - 1\right]\left[\frac{n}{1-n} - 2\right] \frac{1}{1 \times 2 \times 3} + \&c.$$

$$= 1 - \frac{n}{2} + \frac{n}{4}\left(\frac{2n-1}{1 \times 2}\right) - \frac{n}{8}(2n-1)\left(\frac{3n-2}{1 \times 2 \times 3}\right) + \frac{n}{16}$$

$$\left(\frac{2n-1}{1 \times 2 \times 3 \times 4}\right)(3n-2)(4n-3) + \&c.$$

Put $n = 1$, then the series becomes

$$(1 + Z)^a = 1 - \frac{1}{1 \times 2} + \frac{1}{2^2} \cdot \frac{1}{1 \times 2} - \frac{1}{2^3} \frac{1}{1 \times 2 \times 3} + \frac{1}{2^4} \cdot$$

$$\frac{1}{1 \times 2 \times 3 \times 4} + \&c.$$

$$\text{But } e^x = 1 + x + \frac{x^2}{1 \times 2} + \frac{x^3}{1 \times 2 \times 3} + \frac{x^4}{1 \times 2 \times 3 \times 4} + \&c.$$

where $e$ is the base of the Napierian logarithms

Put $x = -\frac{1}{2}$, and we obtain

$$\frac{1}{\sqrt{e}} = e^{-\frac{1}{2}} = 1 - \frac{1}{2} + \frac{1}{2^2} \cdot \frac{1}{1 \times 2} - \frac{1}{2^3} \cdot \frac{1}{1 \times 2 \times 3} + \&c.$$

hence $\frac{1}{\sqrt{e}} = (1 + Z)^a$ when $n = 1$

$$\text{and therefore } \frac{1}{\sqrt{e}} = \left(\frac{n+1}{2}\right)^{\frac{n}{1-n}} = \frac{P_2}{P_1} = \cdot 6. \quad . \quad . \quad (21)$$

In Napier's experiments it was found that the peculiarity already alluded to existed when the pressure outside the orifice was *equal to or less than* $\cdot 6\, P_1$. ; that is, when the pressure outside the orifice was equal to or less than $\frac{3}{5} P_1$, the maximum weight of steam possible, with the initial pressure $P_1$, flowed from the orifice.

To obtain the actual velocity of the steam for maximum weight flow, we find by referring to equations (14) to (17), that

$$v_2 = \sqrt{2\,g\,\frac{n}{n+1}\,\frac{P_1}{\rho_1}} \quad . \quad . \quad . \quad . \quad . \quad . \quad (22)$$

$$\text{or} = \sqrt{2\,g\,\frac{n}{n+1}\,P_1\,V_1} \quad . \quad . \quad . \quad . \quad . \quad . \quad (23)$$

The choice of equations (22) and (23) depends upon whether you have at your command tables of the densities or the volumes of steam at different pressures.

It may be of use to give the value of the several coefficients in equations (19), (20), (22), in tabular form for the various values of the exponent $n$. These will be found in Table I., given below.

TABLE I.

| The different known values of $n$. | $\frac{P_2}{P_1}$ in equation (20). | Co-efficient of $A_2\sqrt{P_1\,\rho_1}$ in equation (19). | $\sqrt{2g\frac{n}{n+1}}$ coefficient of $\sqrt{\frac{P_1}{\rho_1}}$ in equations (22) and (23). | $n\frac{P_2}{P_1}$ coefficient of $P_1\,A_2$ in equation of impulse. It is column 1 multiplied by column 2. |
|---|---|---|---|---|
| 1·4 | ·528 | 3·88 | 6·14 | ·735 |
| 1·3 | ·546 | 3·79 | 6·04 | ·714 |
| 1·135 | ·577 | 3·60 | 5·85 | ·645 |
| $\frac{10}{9}$ | ·582 | 3·57 | 5·82 | ·646 |
| 1 | ·6 | 3·40 | 5·75 | ·600 |

## CHAPTER II.

### IMPULSE OF THE STEAM JET.

HAVING obtained the maximum weight of steam flowing out of the orifice $A_2$ per second under a given boiler pressure $P_1$, equation (19), also the velocity of flow from the orifice, equation (22) or (23), we can easily find the momentum generated per second in the steam jet. As the impulse equals the momentum generated, we have—

Impulse per second of steam jet = momentum generated per second = mass of steam flowing out of orifice per second × its velocity $= \frac{Q\,\text{max.}\,v_2}{g}$.

$$= \frac{A_2}{g}\left(\frac{2}{n+1}\right)^{\frac{1}{n-1}} \sqrt{2\,g.\frac{n}{n+1}.P_1\,\rho_1} \times \sqrt{2g\,\frac{n}{n+1}\,\frac{P_1}{\rho_1}}$$

$$= P_1\,A_2\left(\frac{2}{n+1}\right)^{\frac{1}{n-1}} \times \frac{2\,n}{n+1}.$$

$$= P_1\,A_2\,n\left(\frac{2}{n+1}\right)^{\frac{n}{n-1}}$$

$$= P_1\,A_2\,n\frac{P_2}{P_1} \text{ or } n\,P_2\,A_2 \quad . \quad . \quad . \quad . \quad . \quad (24)$$

The coefficient $\frac{n\,P_2}{P_1}$ of $P_1\,A_2$ in this last equation will have different values according to the value given to $n$. In the last column of Table I. will be found the actual numerical value of this coefficient for each of the usual values of $n$.

An equation similar to equation (24) has previously been obtained by Professor Elliott, and published in a paper by him on a "Hydro Mechanical Theorem," read before the Edinburgh Mathematical Society. The coefficient of $P_1\,A_2$ in his paper is given as

$$2\left(\frac{1}{n}\right)^{\frac{1}{n-1}}\,{}^{*}$$

obtained by writing down the equation of the impulse in

---

* Impulse under any conditions

$$= \frac{\text{weight flow}}{g} \times \text{velocity}$$

$$= \frac{A_2\,\rho_2}{g}\,v_2{}^2 = A_2\,\rho_2 \times 2\,g\,\frac{n}{n-1}\left(\frac{P_1}{\rho_1} - \frac{P_2}{\rho_2}\right) \textit{ vide eq.}\ (15).$$

This will be a maximum when $\frac{P_1}{\rho_1}\,\rho_2 - P_2$ is a maximum.

Differentiating, and remembering that $P = K\,\rho^n$, and $\frac{P_1}{\rho_1}$ a constant, we get—

$$\frac{P_2}{\rho_2} = \frac{1}{n}\,\frac{P_1}{\rho_1} \text{ and } \therefore \frac{P_2}{P_1} = \frac{1}{n}\,\frac{\rho_2}{\rho_1^n}.$$

*For continuation of footnote see next page.*

terms of pressures and densities, and then finding the relation between them when the impulse is a maximum. This is then taken as the probable value of the impulse for the steady state. The experiments of Napier, and the subsequent experiments of Mr. B. G. Buttolph,† show that when steam flows from an orifice, and the pressure in the space into which the steam is flowing is approximately ·6 times the pressure in the boiler, then, on lowering the pressure in the space outside the exit orifice, the flow in pounds per second remains almost unaltered, and the flow is a maximum. This will be gathered from Table II., on page 16.

This table also shows that if the ratio of the pressures is greater than ·6 the flow in pounds per minute falls off. This has been indicated in the analysis. The ratio of

---

Putting in the value for $\frac{P_2}{\rho_2}$ in the equation of impulse, we get—

$$\text{Maximum impulse} = 2\,P_1\,A_2\,\frac{\rho_2}{\rho_1}.$$

Now, from equation $P = K\,\rho^n$ we get—

$$\left(\frac{\rho_2}{\rho_1}\right)^n = \frac{P_2}{P_1}, \text{ and from above } = \frac{1}{n}\,\frac{\rho_2}{\rho_1}$$

from which we obtain

$$\frac{\rho_2}{\rho_1} = \left(\frac{1}{n}\right)^{\frac{1}{n-1}}$$

Putting this in the last equation of maximum impulse, we find—

$$\text{Maximum impulse} = 2\,P_1\,A_2\left(\frac{1}{n}\right)^{\frac{1}{n-1}}$$

This may be written as—

$$2\,P_2\,A_2\,.\,\frac{P_1}{P_2}\,.\left(\frac{1}{n}\right)^{\frac{1}{n-1}} = 2\,P_2\,A_2\,.\,n\,\frac{\rho_1}{\rho_2}\left(\frac{1}{n}\right)^{\frac{1}{n-1}}$$

$$= 2\,P_2\,A_2\,n\,\frac{\left(\frac{1}{n}\right)^{\frac{1}{n-1}}}{\left(\frac{1}{n}\right)^{\frac{1}{n-1}}} = 2\,n\,P_2\,A_2,$$

the quantity $P_2$ in this equation differing in value from that given in equation (24). One is nearly twice the other.

The greater portion of this footnote has been taken from Professor Elliott's paper.

---

† Proceedings of the American Society of Mechanical Engineers, 1888.

pressures obtained by Professor Elliott for *maximum impulse* (not maximum weight flow) is—

$$\left(\frac{1}{n}\right)^{\frac{n}{n-1}}$$

and the impulse obtained with this ratio is—

$$2n\, P_2\, A_2 \text{ or } 2\left(\frac{1}{n}\right)^{\frac{1}{n-1}} P_1\, A_2.$$

This expression gives a slightly greater impulse than equation (24), to the extent of about 4 or 5 per cent; but the numerical value found from equation (24) is the one that will be used in the following pages, because the results of experiments with injectors appear to show a nearer coincidence with those results predicted on the assumption that the pressure in the steam jet orifice is as nearly as possible ·6 times the boiler pressure. This coefficient (·6) is the ratio of the pressures giving maximum weight flow when the exponent $n$ of the equation,

$$P\, V^n = \text{constant},$$

TABLE II.—*Experiments of Mr. Buttolph. Results of Experiments on the flow of saturated steam through a tube ·275in. internal diameter and 8in. long, rounded at the entrance to reduce contraction.*

| Absolute pressure in pounds per square inch. | | Difference of pressure in pounds per square inch. | Flow of steam in pounds per minute. | Ratio of *press. at exit* to pressure at entrance. |
|---|---|---|---|---|
| At the entrance to pipe. | In space at exit of steam from pipe. | | | |
| 83·8 | 19·1 | 64·7 | 3·82 | ·23 |
| 84·3 | 24·4 | 59·9 | 3·84 | ·29 |
| 86·0 | 29·5 | 56·5 | 4·02 | ·34 |
| 83·8 | 34·1 | 49·7 | 3·86 | ·41 |
| 85·7 | 39·1 | 46·6 | 3·90 | ·46 |
| 85·0 | 43·8 | 41·2 | 3·82 | ·51 |
| 86·8 | 48·9 | 37·9 | 3·86 | ·56 |
| 86·8 | 54·2 | 32·6 | 3·68 | ·62 |
| 86·3 | 58·9 | 27·4 | 3·56 | ·68 |

is equal to unity [*vide* equation (21)]. It would seem probable that results calculated on this assumption would be near the truth, when we come to consider that steam, as found in the boiler, is rarely quite dry; and, unless it is quite dry, there will be the mass of moisture to be carried along by the steam, and, consequently, the velocity thereof would be slightly reduced. The exponent $\frac{10}{9}$, proposed by Rankine for adiabatic expansion of steam, according to Zeuner, only holds when the steam is *quite dry* at the commencement of expansion, being less for wet steam. Again, if we look at the numbers in columns 3, 4, and 5 of Table I., we see at a glance that the nearer steam approaches to a so-called perfect gas, the greater becomes its velocity, its weight flow, and its impluse.

It will be found further on that whichever value of the impulse of the steam jet is selected, it leaves the ratio of the different parts of the injector unaffected; and this is the chief use of the investigation.

It is often convenient to have the several quantities used in calculations placed consecutively in the form of a table. This has been done in Table III., given below. The first column gives the absolute pressures per square inch of boiler steam. In the second column will be found the volume occupied by 1lb. of dry saturated steam at the pressure given in column 1. The product of the pressure per square foot, and the volume in cubic feet of 1lb. of dry saturated steam, is given in the third column, while the density—weight per unit of volume—is found in column 4. The density is numerically the same as one divided by the volume in cubic feet of 1lb. of dry saturated steam found in the second column. The temperature corresponding to the pressure found in column 1 is given in the fifth column in degrees on the Fahrenheit scale. The true sensible heat of 1lb. of water, at the temperature given in the fifth column, will be found in column 6. It will be found that these values are slighly above those given by the approximate expression $(t - 32)$. The total latent heat of the same quantity of steam is given in the next column. After these quantities will be found the calculated values of the velocity of steam issuing from an injector steam cone when the maximum weight flow occurs. This has been taken as a fair value of the actual velocity in practice. In the equation for the maximum weight flow will be found the product $\sqrt{P\rho}$. The value of this product is given in the eleventh column, and it will, no doubt, be found of some

## TABLE III.

| Boiler pressure per sq. inch $= \frac{P_1}{144}$. | Volume of 1lb. of steam at a pressure $P_1$ lbs. per square foot. $= V_1$ cubic feet. | Product of the pressure per sq. foot into the volume of 1lb. steam in cubic feet. $= P_1V_1$ foot-pounds. | Density of the steam at a pressure of $P_1$ lbs. per square foot—*i.e.*, weight of 1 cubic foot steam. $= \rho_1$. | Temperature of steam in degrees Fah., corresponding to a pressure of $P_1$ lbs. per square foot. $= t_1°$. | Sensible heat of 1lb. steam at temperature $t_1°$—*i.e.*, units of heat required to raise the temp. of 1lb. of water from 32° F. to $t°_1$ F. $= h_1$. | Latent heat of 1lb. of steam at a temp. $t°$ F.—*i.e.*, units of heat required to convert 1lb. of water at $t_1°$ F. into 1lb. steam at $t_1°$ F. = L. | Velocity of steam jet calculated by equation (10). | Velocity of steam jet calculated by aid of equation (22) or (23) when $n = 1$. | Velocity of steam jet calculated by aid of equation (22) or (23) when $n = \frac{10}{9}$. | Square root of the product of pressure per square foot into the density $= \sqrt{P\rho}$ for equati'n of weight flow. | Product ·6 $P_1 = P_2$ for equation of impulse. |
|---|---|---|---|---|---|---|---|---|---|---|---|
| 1 | 330·4 | 47577 | ·0030 | 102·0 | 70·04 | 1043·01 | .. | 1238 | .. | ·6 | 86 |
| 2 | 171·9 | 49507 | ·0058 | 126·3 | 94·37 | 1026·09 | .. | 1262 | .. | 1·3 | 173 |
| 3 | 117·3 | 50673 | ·0085 | 141·6 | 109·76 | 1015·38 | .. | 1277 | .. | 1·9 | 259 |
| 4 | 89·51 | 51734 | ·0112 | 153·1 | 121·27 | 1007·37 | .. | 1291 | .. | 2·5 | 345 |
| 5 | 72·56 | 52387 | ·0138 | 162·4 | 130·56 | 1000·90 | 1292 | 1299 | 1332 | 3·1 | 432 |
| 10 | 37·83 | 54561 | ·0264 | 193·3 | 161·66 | 979·23 | 1321 | 1326 | 1360 | 6·2 | 864 |
| 15 | 25·85 | 56097 | ·0386 | 213·0 | 181·60 | 965·32 | .. | .. | .. | 9·2 | 1296 |
| 20 | 19·74 | 56993 | ·0506 | 227·9 | 196·65 | 954·81 | .. | .. | .. | 12·1 | 1728 |
| 50 | 8·338 | 60565 | ·1199 | 280·9 | 250 35 | 917·26 | 1404 | 1398 | 1426 | 29·4 | 4320 |
| 70 | 6·076 | 61542 | ·1645 | 302·7 | 272·65 | 901·63 | .. | .. | .. | 40·1 | 6048 |
| 80 | 5·358 | 61719 | ·1866 | 311 8 | 281·95 | 895·11 | . | .. | .. | 44·8 | 6912 |
| 90 | 4·796 | 62156 | ·2085 | 320·1 | 290·37 | 889·19 | .. | .. | .. | 52·0 | 7776 |
| 100 | 4·342 | 62737 | ·2302 | 327·6 | 298·09 | 883·77 | 1408 | 1420 | 1457 | 58·0 | 8640 |
| 110 | 3·969 | 62885 | ·2519 | 334 6 | 305·24 | 878·74 | .. | .. | .. | 63·2 | 8504 |
| 120 | 3·656 | 63177 | ·2735 | 341·0 | 311·88 | 874·07 | .. | .. | .. | 69·0 | 10368 |
| 130 | 3·390 | 63461 | ·2949 | 347·1 | 318·12 | 869·68 | .. | .. | .. | 74·0 | 11234 |
| 140 | 3·161 | 63705 | ·3163 | 352·8 | 324·00 | 865·55 | .. | .. | .. | 80·0 | 12096 |
| 150 | 2·962 | 63979 | ·3376 | 358·2 | 329·56 | 861·63 | 1414 | 1434 | 1465 | 86·0 | 12960 |
| 160 | 2·786 | 64180 | ·3588 | 363·3 | 324·85 | 857·91 | .. | .. | .. | 91·0 | 13824 |
| 170 | 2·631 | 64382 | ·3800 | 368·2 | 339·89 | 854·36 | .. | .. | .. | 96·0 | 14688 |
| 180 | 2·493 | 64620 | ·4012 | 372·8 | 344·70 | 850·96 | .. | .. | .. | 102·2 | 15552 |
| 190 | 2·368 | 64788 | ·4223 | 377·3 | 349·33 | 847·70 | .. | .. | .. | 108·0 | 16416 |
| 200 | 2·256 | 65525 | ·4433 | 381·6 | 353·76 | 844·57 | 1432 | 1452 | 1483 | 112·1 | 17280 |
| 250 | 1·825 | 65700 | ·5478 | 400·8 | 373·70 | 830·45 | .. | 1454 | 1485 | 140·4 | 21600 |
| 300 | 1·535 | 66812 | ·6515 | 417·3 | 390·93 | 818·30 | .. | 1461 | 1492 | 168·5 | 25920 |
| 500 | ·942 | 67824 | 1·0617 | 467·4 | 443·50 | 781·02 | .. | 1477 | 1509 | 277·5 | 43200 |

use in calculating results when necessary. The product $\cdot 6 P_1 = P_2$ for the equation of impulse is given in the last column.

The table has been formed because the quantities found therein are not always at hand when required; and as they are very necessary when reading the matter contained in these few pages, it has been thought not out of place to give it here.

---

## CHAPTER III.

### Approximate Calculations for Injector.

Having now obtained all the necessary preliminary quantities (such as velocity of steam in the steam cone, the weight flow of steam, and the impulse of the jet), we will devote a little space to an exceedingly neat and simple method of approximating very closely to the results of an injector's working, before passing on to the general equation of impulse and momentum.

This method is to be recommended especially to those who have an elementary knowledge of the "principles of mechanics," and at the same time are not familiar with the elements of the differential and integral calculus. The only item which of itself may be troublesome to find is the velocity of the steam jet; but this may be obtained by the aid of equation (10), the getting of which only requires very elementary mathematics. Hence, if preferable, the matter between equation (10) and this paragraph may be omitted, and an exceedingly near result obtained therewith.

Let $\beta$ = the number of pounds of cold feed water supplied to the injector per pound of steam; then the total delivery per pound of steam = 1lb. steam + $\beta$ lbs. of water = $(1+\beta)$ lbs. of water as it enters the boiler. Now, the total heat in 1lb. of steam, reckoned from 32 deg. Fah., with the same notation as before $= h_1 + x_1 L_1$ thermal units; and if the suffix 3 denote the different quantities in the combining cone orifice, also suffix 5 those in the cold water supply pipe or chamber, then the total heat in 1lb. of steam, after condensation in the combination cone, is $h_3$; and heat given up by 1lb. steam in condensation

$$= h_1 + x_1 L_1 - h_3 \text{ thermal units.}$$

Also heat gained by $\beta$ lbs. of cold water in condensing 1lb. of steam

$$= \beta (h_3 - h_5) \text{ thermal units;}$$

and as the heat lost by the steam must be that gained by the cold water, we have—

$$h_1 + x_1 L_1 - h_3 = \beta (h_3 - h_5)$$

or

$$\frac{h_1 + x_1 L_1 - h_3}{h_3 - h_5} = \beta \quad . \quad . \quad . \quad . \quad (25)$$

This equation is, of course, based on the assumption that the kinetic energy of the delivery jet on entering the boiler may be neglected. It is assumed also that the overflow is open to the atmosphere, and hence the temperature of the delivery jet cannot be greater than 212 deg. Fah. It is rarely above 170 deg. Fah., although in the best constructed injector (single type) it has been made to reach the high temperature of 190 deg. Fah. With a compound injector—that is, an injector having more than one set of cones—the temperature of the feed water as it enters the boiler may be raised as high as 280 deg. Fah. These injectors will be more fully dealt with in succeeding pages. Let the same symbols be used as before, and assume that every pound of steam carries into the boiler $\beta$ pounds of cold feed water along with it.

It has been previously remarked that, as impulse and momentum are of the same kind or dimensions, we may add them together when necessary. Now, returning for a moment to fig. 1, we have before impact or mixture the momentum of the steam issuing from $A_2$, and the impulse due to the pressure of the atmosphere on the surface of the cold feed water, together with that due to the actual head of feed water itself. This last quantity is a positive pressure when the injector is below the surface of feed water, but a negative pressure, or what is commonly termed a suction, when the feed water is to be raised to the injector. If $Pa$ be the pressure of the atmosphere, and $P_6$ the pressure due to the height, the surface of the cold feed water is situated above the orifice $A_3$. Then the net pressure $P_5$ in the water supply pipe or chamber near the orifice $A_3$ will be

$$Pa + P_6 \text{ or } Pa - P_6$$

according as the water is obtained from a level above or below the injector. So far nothing has been said about the velocity $v_5$ of the incoming water. It will be found that if a number of injectors are examined, the chamber surrounding the steam cone, into which the water flows, is large compared with the orifice $A_3$ of the condensing cone; and therefore the velocity of the water in this chamber will be correspondingly small, and can be neglected. The useful effect of the pressure $P_5$ acting over the orifice $A_3$ on the mixed steam and water will be of the nature of an impulse producing velocity in the mixed steam and water, helping to drive it into the delivery cone across the overflow space.

Outside of the orifice $A_3$ we have a quantity of condensed steam and feed water moving with the velocity it has acquired in passing through the combining cone. Also there is the impulse of the pressure $P_3$ in the overflow chamber on the jet of water as it issues from $A_3$ tending to drive it back or retard its motion. If the overflow is open to the atmosphere, this pressure will be that of the atmosphere. In this particular calculation the overflow is assumed open. Putting these several quantities into numerical form, we find—

Momentum of the steam issuing from the orifice $A_2$ per second

$$= \frac{A_2 \rho_2 v_2}{g} v_2.$$

The areas of all orifices are measured in square feet, while the density of water = weight per cubic foot.

Impulse of pressure $P_5$ on orifice $A_3$ per second

$$= A_3 P_5 \times 1 = A_3 (Pa + P_6) = A_3 (Pa + \rho H),$$

H being the head of cold water, producing the pressure $P_6$ lbs. per square foot.

Momentum of combined steam and water on issuing from $A_3$

$$= \frac{A_3 \rho v_3}{g} v_3,$$

and impulse due to the atmospheric pressure on $A_3$, tending to retard the flow,

$$= A_3 P_3 \times 1 = A_3 Pa.$$

Putting these values in the equation representing the fact that the sum of the impulses and momentum before impact equals the sum of impulses and momentum after impact, we have—

$$\frac{A_2 \rho_2 v_2^2}{g} + A_3 (Pa + \rho H) = \frac{A_3 \rho v_3^2}{g} + A_3 Pa$$

or $$A_2 \rho_2 v_2^2 + g \rho H A_3 = A_3 \rho v_3^2 \quad . \; . \; . \; . \; . \; (26)$$

The weight of combined water and steam delivered per second $= A_3 \rho v_3$ lbs.

Also weight of steam flowing per second $= A_2 \rho_2 v_2$ lbs.; therefore—

$$\frac{A_2 \rho_2 v_2}{A_3 \rho \ v_3} = \frac{\text{steam supplied}}{\text{water and steam delivered to boiler}} = \frac{1}{1+\beta} \quad . \quad . \quad (27)$$

Now, divide both sides of equation (26) by $A_3 \rho v_3$, and we get—

$$\frac{A_2 \rho_2 v_2}{A_3 \rho \ v_3} v_2 + \frac{g \rho H A_3}{A_3 \rho v_3} = \frac{A_3 \rho v_3}{A_3 \rho v_3} . v_3.$$

Substituting from equation (27), we get—

$$\frac{1}{1+\beta} . v_2 + \frac{g H}{v_3} = v_3,$$

and

$$v_2 = (1+\beta) \left[\frac{v_3{}^2 - g H}{v_3}\right] \quad . \quad . \quad . \quad (28)$$

Or, solving for $v_3$ instead—

$$v_3 = \frac{v_2}{2(1+\beta)} \pm \sqrt{g H + \frac{v_2{}^2}{4(1+\beta)^2}}$$

The lift cannot be great, so that the second term of the right-hand side must always have the positive sign in front of it, so that the velocity $v_3$ may be large enough to enter the boiler. When H is positive, the cold water is assisting the injector; but when negative it is acting against the injector. The above equation may then be written as—

$$v_3 = \frac{v_2}{2(1+\beta)} + \sqrt{\frac{v_2{}^2}{4(1+\beta)^2} \pm g H} \quad . \quad . \quad (29)$$

If the cold feed water is on a level with the injector, $H = O$, and equation (29) becomes

$$v_3 = \frac{v_2}{1+\beta} \quad . \quad . \quad . \quad . \quad . \quad . \quad . \quad . \quad (30)$$

which is exactly what we should have expected, because the momentum of 1lb. of steam before impact must equal the momentum of $(1+\beta)$ pounds of delivery water after impact.

The above reasoning to establish equation (28) may be stated somewhat differently, with the same result in the end; thus the mass of combined water and steam passing through the orifice $A_3$ per second

$$= \frac{A_3 v_3 \rho}{g}.$$

The total pressure on the orifice $A_3$ produced by the head of cold water H

$$= \text{H} \rho \text{A}_3 \text{lb.}$$

This force, acting for one second, will produce a velocity $v$ in the above mass, given by the equation—

$$\text{Force} \times \text{time} = \text{mass} \times \text{velocity};$$

or

$$v = \frac{\text{H} \rho \text{A}_3}{\frac{\text{A}_3 v_3 \rho}{g}} = \frac{g \text{H}}{v_3}$$

This is the equivalent velocity of the combined steam and water at $A_3$, due to the head of cold water H. Hence the equivalent momentum

$$= \frac{(1 + \beta)}{g} \frac{g \text{H}}{v_3}$$

when considering the flow of 1lb. steam and $\beta$lb. of cold water. Now, because the momentum before impact = momentum after impact, we have—

Momentum of steam + equivalent momentum of feed water = momentum of combined steam and feed water on issuing from orifice $A_3$;

or

$$\frac{1}{g} \times v_2 + (1 + \beta) \frac{\text{H}}{v_3} = \frac{(1 + \beta)}{g} v_3,$$

which simplifies to—

$$v_2 = (1 + \beta) \left[ \frac{v_3{}^2 - g \text{H}}{v^3} \right].$$

This is a repetition of equation (28).

---

## CHAPTER IV.

### Sizes of Cones.

Assume that the injector is required to supply Mlb. of cold feed water per hour to a boiler, and it is required to find the diameters of the orifices, or, in other words, the size of the injector to perform this amount of work. For every $\beta$lb. of water supplied to the boiler there is 1lb. of steam required; hence the weight of steam used per hour is $\frac{M}{\beta}$lb., and the weight per second

$$= \frac{M}{3{,}600\,\beta}\text{lb.}$$

The volume of this weight of steam

$$= \frac{M}{3{,}600\,\beta} \times \text{volume of 1lb. steam at pressure } P_2$$

$$= \frac{M}{3{,}600\,\beta\,\rho_2} \text{ or } \frac{M\,V_2}{3{,}600\,\beta}.$$

But volume of steam passing through orifice $A_2$ per second $= A_2\,v_2$ cubic feet. Therefore—

$$A_2\,v_2 = \frac{M}{3{,}600\,\beta\,\rho_2} = \frac{M\,V_2}{3{,}600\,\beta}.$$

Inserting the numerical value of $v_2$, previously found either from equation (10) or (22), and substituting $\pi\,r_2^2$ for the area $A_2$ (the radius being $r_2$), we find that—

$$r_2 = \cdot0094\sqrt{\frac{M\,V_2}{\beta\,V_2}}\text{ft.,}$$

or

$$= \cdot113\sqrt{\frac{M\,V_2}{\beta\,v_2}}\text{in.,} \quad . \quad . \quad . \quad . \quad . \quad (31)$$

The combining cone orifice must be made large enough to allow of the passage of the steam and feed water—*i.e.* $(1+\beta)$lb. of water for each pound of steam, and therefore $\frac{1+\beta}{\beta}$lb. for each pound of cold feed water. Hence there will be—

$$\frac{1+\beta}{\beta}\text{Mlb.}$$

of water delivered through the orifice $A_3$ per hour, with a velocity $v_3$ given by equation (29). Now, if $A_3$ is the area of the orifice, and $r_3$ its radius, the weight of water passing per second

$$= v_3\,A_3\,\rho = \pi\,r_3^2\,v_3\,\rho\text{lb.} = \frac{1+\beta}{\beta} \times \frac{M}{3{,}600}$$

hence $$r_3 = \cdot0094\sqrt{\frac{(1+\beta)}{\beta}\,\frac{M}{\rho\,v_3}}\text{ft.,}$$

or $$= \cdot 113 \sqrt{\frac{(1+\beta)}{\beta} \frac{M}{\rho\, v_3}} \text{ in.}, \quad \ldots \ldots \quad (32)$$

In can be probably made slightly larger than this, because the whole of the steam may not have been condensed by the time it reaches this orifice, should the cone be rather short. This being so, the entrance to the diverging cone will be slightly smaller than the orifice $A_3$.

As an example of the utility of the above equations, let it be required to ascertain the sizes of the orifices of an injector which is to supply a boiler with 10,000lb. of water per hour (this is equivalent to a boiler of 100 horse power). The temperature of the water source is assumed to be 60 deg. Fah., the lift to be 10ft., and the temperature of the water as it enters the boiler 170 deg. Fah., the pressure (absolute) of the steam in the boiler being 100lb., the steam being dry.

The quantity of feed water per pound of steam used is given by equation (25) to be—

$$\beta = \frac{298{\cdot}09 + 883{\cdot}77 - 138{\cdot}35}{138{\cdot}35 - 28} = 8{\cdot}82\text{lb.}$$

This calculation is facilitated by the use of Table III. The velocity of the steam as it issues from the steam orifice can be obtained by the aid of equation (10), thus—

$$\frac{v_2^{\,2}}{2g} = 144 \times {\cdot}016\,(100 - 60) + 772\left[298{\cdot}09 - 262{\cdot}25 + \frac{883{\cdot}77}{788{\cdot}62} \cdot 35{\cdot}05 - (753{\cdot}57 \times {\cdot}0488)\right]$$

$$= 82 + 772\,(35{\cdot}84 + 39{\cdot}2 - 35{\cdot}2)$$

$$v_2 = \sqrt{64{\cdot}4 + 30{,}799} = 1{,}408\text{ft. per second.}$$

The accuracy of the above result depends to a very great extent upon the accurate finding of the Napierian logarithm of the ratio of the absolute temperatures. As tables of these are not generally at hand, it is better perhaps to use equation (22) or (23), to obtain the same result. The exponent $n$ we shall take as unity, this value giving results very near the truth; we then get

$$v_2 = \sqrt{2g \times \tfrac{1}{2} \times 144 \times 100 \times 4{\cdot}342} = 1{,}420\text{ft. per second.}$$

The velocity of the delivery water in the orifice $A_3$ is found by the aid of equation (29); therefore

$$v_3 = \frac{1420}{19{\cdot}64} + \sqrt{\left(\frac{1420}{19{\cdot}64}\right)^2 - 32{\cdot}2 \times 10}$$
$$= 143{\cdot}5\text{ft. per second.}$$

The radius of the steam orifice is from equation (31)—

$$r_2 = {\cdot}113\sqrt{\frac{10{,}000 \times 7{\cdot}024}{8{\cdot}82 \times 1{,}420}}$$
$$= {\cdot}27\text{in.}$$

The radius of the combining cone orifice is obtained in a similar manner from equation (32), thus—

$$r_3 = {\cdot}113\sqrt{\frac{9{\cdot}82 \times 10{,}000}{8{\cdot}82 \times 62{\cdot}5 \times 143{\cdot}5}}$$
$$= {\cdot}13\text{in.}$$

If, in equation (25), we substitute the approximate values, $h_1 = t_1 - 32$, $L_1 = 1{,}114 - {\cdot}7\,t_1$, $h_3 = t_3 - 32$, and $h_5 = t_5 - 32$, we find that

$$\beta = \frac{1{,}114 + {\cdot}3\,t_1 - t_3}{t_3 - t_5}$$

Inserting this result in equation (30), we find that

$$v_3 = \frac{v_2}{1 + \dfrac{1{,}114 + {\cdot}3\,t_1 - t3}{t_3 - t_5}} = \frac{v_2\,(t_3 - t_5)}{1{,}114 + {\cdot}3\,t_1 - t_5} \quad . \quad . \quad . \quad (33)$$

For a given boiler pressure, $v_2$ and $t_1$ are constant; hence the velocity of the jet as it enters the diverging cone is directly proportional to its temperature, and therefore the injector will feed against a higher pressure according as the temperature of delivery is increased; but as a high temperature corresponds to a low value for $\beta$, then we may conclude that a maximum temperature of delivery will necessitate the minimum amount of cold feed water supplied per pound of steam.

Again, to ascertain the influence of the temperature of the feed water on the working of the injector, we must find the rate of increase of $v_3$ with respect to $t_5$. This may be done by inspection of the last equation. In the denominator $t_5$ will always be small compared with the remainder, and for approximate purposes we may neglect it. We then see that $v_3$ increases directly as $t_5$ is lowered, $t_3$ being constant;

at the same time, if $t_5$ is constant, then $v_3$ increases directly as $t_3$; and therefore, when forcing the feed against a high boiler pressure, it is advantageous to raise the temperature of the feed before it enters the main injector by means of an exhaust injector. This will be dealt with in succeeding pages. This will be more forcibly seen if the first differential coefficient of $v_3$ with respect to $t$ be found. The last equation also shows that the velocity $v_3$ varies approximately directly as the rise of temperature, $t_3 - t_5$, of the feed water. If the value of $\beta$, just found, be placed in equation (29), we find, in addition to the above, that the temperature $t_3$ of the delivery increases directly with the height through which the water is lifted to the injector. As this cannot exceed about 25ft., the increase in temperature due to this cannot be a large amount.

Following out the same reasoning with equations (31) and (32), we see that the diameters of the steam and delivery nozzles vary approximately directly as the square root of the rise in temperature of the feed water.

---

## CHAPTER V.

### LIMITS TO WORKING OF INJECTOR.

The limits between which the injector will work can also be determined in like manner, for the head due to the velocity $v_3$ must not be less than that due to the pressure in the boiler; hence $\frac{v_3^2}{2g}$ cannot be less than $\frac{P_1}{\rho}$ or that $v_3$ cannot be less than $\sqrt{\frac{2gP_1}{\rho}}$, otherwise the jet of water would not be able to force back the water which is tending to flow out of the boiler.

For $v_3$ above, we can substitute its approximate value in equation (33), and for $v_2$ in that equation write its equivalent in equation (22). We then see that the rise in temperature $(t_3 - t_5)$ of the feed water cannot be less than

$$(1{,}114 + \cdot 3\,t_1 - t_5)\sqrt{\frac{\rho_1}{\rho}}. \quad . \quad . \quad . \quad . \quad . \quad . \quad . \quad (34)$$

This quantity, it will be observed, increases with the boiler pressure (because it increases with the boiler temperature and the square root of the density of the steam), and decreases slightly with an increase in the temperature of the cold feed water. Assuming $t_5$ to be 60 deg. Fah., and the boiler pressure 100lb. per square inch, then the above shows that the minimum rise in temperature of the feed water is about 70 deg. Fah. The maximum rise in temperature of the feed water depends upon the capacity of the water to condense the steam in the combining cone, without permitting any overflow of steam or water. As the jet is in contact with the atmosphere generally before entering the diverging cone, its temperature is assumed not to exceed 212 deg., and in practice hardly ever reaches 200 deg. with a single injector. Then, as the maximum range of temperature of feed water corresponds to the minimum amount of water per pound of steam, this maximum temperature of delivery—and therefore maximum range in temperature—of feed water corresponds to the minimum of water per pound of steam. In practice it is advantageous to feed the boiler continually, rather than at intervals, on account of the severe reduction of steam pressure when fed too quickly; this necessitates an injector which will just replace the water as the steam is drawn off. Again, the life of a boiler is considerably lengthened by feeding it with water at a maximum temperature; therefore, on this account also, it is conducive to economy to use an injector only *just* large enough to do the work. This becomes much more apparent when dealing with an exhaust injector, though the extra increase in economy is due to other causes as well. Again, the higher the temperature of the water delivered to the boiler, the greater is the velocity of delivery. A jet of water injected into a boiler with a high velocity will materially augment the circulation of the water in the boiler, and therefore allow it to make steam more easily.

It has been remarked that water cannot enter the boiler if its velocity in the combining cone should be less than that which is required to overcome the head of water due to the pressure of the boiler. This velocity is given by the equation—

$$\frac{v_3{}^2}{2g} = \frac{P_4}{\rho} + \frac{v_4{}^2}{2g} = \frac{P_4}{\rho} = \frac{P_1}{\rho}$$

when the pressure in the overflow chamber is neglected. If $P_3$ represent this pressure, which is assumed to be the pressure in the jet of water as it passes through the overflow gap, then the above equation becomes—

$$\frac{P_3}{\rho} + \frac{v_3{}^2}{2g} = \frac{P_1}{\rho}.$$

The velocity of the jet as it enters the boiler is so small generally that it can be neglected in the above equations.

If we now substitute for $v_3$ in equation (30) its value in the above equations, we obtain two expressions containing $\beta$.

$$\frac{v_2}{1 + \beta} = v_3 = \sqrt{\frac{2g(P_1 - P_3)}{\rho}}$$

and

$$\frac{v_2}{1 + \beta} = v_3 = \sqrt{\frac{2g P_1}{\rho}}$$

according as the pressure in the overflow chamber is taken into account or neglected.

Now, if $v_3$ is a minimum, therefore $\frac{v_2}{1 + \beta}$ must also be a minimum ; and for this condition $\beta$ must be a maximum for any particular pressure in the boiler. Therefore, the value found by solving for $\beta$ must be the greatest amount of feed water per pound of steam that can be taken into the boiler by the steam against the steam pressure $P_1$. If we replace $v_2$ by its equivalent from equation (22), and assume that $n = 1$, we find that

$$\beta \text{ maximum} = \sqrt{\frac{P_1 \rho}{2 \rho_1 (P_1 - P_3)}} - 1 \quad . \quad . \quad . \quad (35)$$

or

$$\beta \text{ maximum} = \sqrt{\frac{\rho}{2 \rho_1}} - 1,$$

according as $P_3$ is taken into account or neglected.

The different values of $\beta$ maximum for several pressures will be found in Table IV., column II. These values at the best can only be taken as approximations, for, in the first place, we have assumed that the pressure in the jet is exactly the same as that in the overflow chamber. It cannot well be less, though it may, the author thinks, be slightly greater, when the overflow gap is small. With a closed overflow, the pressure may be anything up to the pressure upon the overflow valve.

For perfect condensation in the combining cone, we gather from equation (25) that

$$(h_3 - h_5)\beta = h_1 + x_1 L_1 - h_3.$$

The right-hand side of this equation represents the total heat in 1lb. of steam above $t_3$ deg. Fah., and therefore, for that particular boiler pressure, is constant. Replace $\beta$ by its maximum value found in the last equation. Now, as $\beta$

is a maximum, therefore $(h_3 - h_5)$ is a minimum. But $h_3$ is constant, therefore $h_5$ must be a maximum. Hence, if we solve for $h_5$ or $t_5$, we shall obtain the maximum temperature at which the feed water can be supplied to the injector for a given steam pressure and a given temperature of delivery. Thus we find—

$$t_5 \text{ maximum} = t_3 - \frac{1{,}114 + \cdot 3\, t_1 - t_3}{\beta} \quad . \quad . \quad . \quad . \quad (36)$$

The result of this equation, for different boiler temperatures, is shown in the third column of Table IV., while the corresponding rise in temperature of the feed water is to be found in the adjoining column. The temperature of delivery is assumed as 210 deg. Fah. The rise of temperature is, of course, a minimum, because the injector is then delivering its maximum amount of feed water. A smaller quantity of feed would necessitate a greater range of temperature. It may be of interest to find the smallest value $\beta$ for a given rise in temperature of the feed water. Assume that the feed is supplied at 60 deg. Fah., and that it is delivered to the boiler at 210 deg. Fah., we shall then have—

$$\beta \text{ minimum} = \frac{1{,}114 + \cdot 3\, t_1 - t_3}{210 - 60}$$

$$= 6 \cdot 02 + \cdot 002\, t_1 \quad . \quad . \quad . \quad . \quad (37)$$

The several quantities representing $\beta$ minimum will be found in the fifth column of Table IV., and it will be noticed that there is no ratio of feed water to steam so great as 7. If the steam should not be perfectly condensed, it would seem that even slightly less feed water could be used, though in that case the uncondensed steam would produce eddies in the jet, and thus tend to stop the injector. In general these minimum quantities are never reached, and must be regarded as ideal rather than real. When experiments are made with an injector, the minimum value of $\beta$ is usually found for various pressures, but the results obtained are of little practical value. There is one thing to which attention may be drawn in the Table IV. At a pressure of about 250lb. per square inch the maximum and minimum values of $\beta$ appear to coincide, which seems to indicate that a single injector, with the overflow open to the atmosphere, will not work with pressures above 250lb.

This peculiar result is to be inferred from another source. Let it be required to find the maximum pressure P into which the injector will drive the feed water, when the boiler pressure is $P_1$. We have from the equation just previous to number (35)—

$$\left(\frac{v_2}{1+\beta}\right)^2 = v^2{}_3 = \frac{2g\,(P_1 - P_3)}{\rho},$$

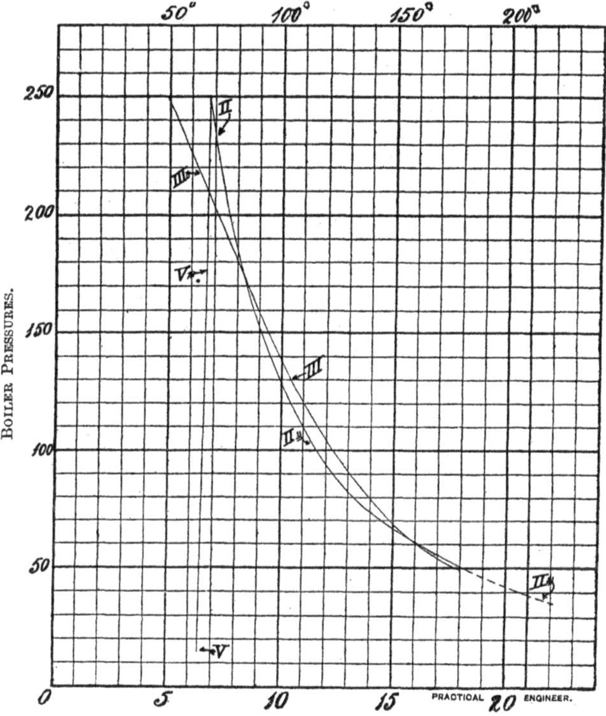


FIG. 3.—POUNDS OF FEED WATER PER POUND OF STEAM.

where $P_1$ is the pressure against which the water is delivered. Call the pressure P, and put in the equivalent of $v_2$ from equation (22). We then find that

$$P = \frac{\rho P_1}{2\rho_1 (1+\beta)^2} + P_3.$$

Here $\beta$ must have its minimum value. If we divide through by 144, we obtain—

$$p = \frac{31{\cdot}25\, p_1}{\rho_1\,(1 + \beta)^2} + p_3.$$

The value of $p$, obtained with $p_1$ = 100lb., is over 200lb.; with $p_1$ = 200lb., $p$ = about 250lb.; and with $p_1$ = 250lb.; $p$ is also about 250lb. So that it would seem useless to

TABLE IV.

| Pressure of boiler steam per square inch. I. | Maximum amount of feed water per pound of steam when considering the pressure in overflow chamber. II. | Maximum temperature of feed water with which injector will work to deliver water into boiler at 210 deg. Fah. III. | Rise in temperature of the feed water in passing through the injector. IV. | Minimum weight of feed water at 60 deg. Fah. when the temperature of delivery is 210 deg. Fah. V. |
|---|---|---|---|---|
| 15 | — | — | — | 6·44 |
| 20 | — | — | — | 6·47 |
| 50 | 18·3 | 180 | 30 | 6·58 |
| 70 | 14·5 | 145 | 65 | 6·62 |
| 80 | 13·4 | 139 | 71 | 6·64 |
| 100 | 11·6 | 124 | 86 | 6·67 |
| 120 | 10 4 | 110 | 100 | 6·70 |
| 140 | 9·6 | 100 | 110 | 6·72 |
| 160 | 8·8 | 90 | 120 | 6·74 |
| 180 | 8·2 | 79 | 131 | 6·76 |
| 200 | 7·7 | 72 | 138 | 6·78 |
| 250 | 6·8 | 50 | 160 | 6·82 |

attempt to feed a boiler with water at a higher pressure than 250lb. per square inch with an ordinary single injector; in fact, it could only be done at 250lb. with most careful design and accurate workmanship. The above equation is of little use for estimating the maximum pressures, for the same reasons that have previously been enumerated.

In fig. 3 will be found curves plotted to represent the results of Table IV. The number near each curve indicates the number of the column whose results it represents. The abscissæ are boiler pressures, and the ordinates are (II.) the maximum amount of feed water per pound of steam, (III.) maximum temperature of feed water when the temperature of delivery is 210 deg. Fah., (IV) rise in temperature of feed water in passing through the injector.

---

# CHAPTER VI.

## Alternative Calculations.

After obtaining expressions for the velocity, the maximum weight flow, and the impulse of the steam jet, we passed on directly to some *approximate* calculations respecting the ordinary working of the injector. We will now assume that there are certain resistances existing, such as friction and eddy motion, which tend to retard its action; and also that the overflow chamber may or may not be open to the atmosphere. The author has derived great assistance in what immediately follows from a perusal of Professor Elliott's paper, previously referred to, and, in fact, some of the ideas brought forward here have been suggested in that paper.

The law of the "Conservation of Momentum" has already been cited more than once, and, as we require it again, we repeat it once more.

*The sum of the impulses and momenta before impact equals the sum of the impulses and momenta after impact.*

Applying this to the injector, we find that the impulse of the steam jet + the impulse of the head of feed water + the impulse of the atmosphere on the feed water = momentum of the delivery jet + the impulse of the pressure in the overflow chamber + the impulse due to friction, eddies, &c.

The impulse of the steam jet is given in equation (24) as

$$n\,P_2\,A_2 \text{ or } n\,P_1\,A_2\,\frac{P_2}{P_1}.$$

The effect of the atmosphere on the feed water will be equivalent to an extra head of feed water due to the atmospheric pressure.

The water pipe and chamber above the combining cone are generally much larger than is required for feeding the injector with water; therefore in general there will be a static pressure on the annular orifice of the feed-water entrance, as well as the momentum of the feed water, unless the relative sizes of the orifices are very nicely adjusted for the maximum efficiency of the injector.

If we call $P_5$ the pressure due to the combined head and atmosphere in the water orifice, and $A_5$ the area of that orifice, we have for the impulse of the feed water and atmosphere

$$2 P_5 A_5 \text{*}$$

The resistance due to friction, &c., will produce an impulse tending to retard the motion of the steam and water in the combining cone; and it will be proportional to the square of the velocity. Let $f$ be the coefficient of resistance, then the total equivalent pressure over the orifice $A_3$

$$= f \rho \frac{v_3^2}{2g} A_3 = \text{impulse.}$$

The impulse due to the pressure in the jet as it enters the diverging cone will be

$$P_3 A_3.$$

The momentum of the delivery jet per second as it issues from the combining cone orifice is

$$\frac{A_3 v_3 \rho}{g} v_3.$$

Collecting these several terms according to the law of the "Conservation of Momentum," we find that

$$n P_2 A_2 + 2 P_5 A_5 = \frac{A_3 \rho v_3^2}{g} + \frac{A_3 \rho f v_3^2}{2g} + P_3 A_3 \quad . \quad . \quad (38)$$

The pressure in the jet at the combining cone orifice and the delivery cone orifice will be the same while the jet

* Impulse = change of momentum per second.

Let H = head of feed water + atmosphere.

Velocity due to this head = $\sqrt{2 g H}$.

Weight of water flowing through orifice per second = volume × density = $A_5 v_5 \rho = A_5 \rho \sqrt{2 g H}$.

Momentum = $\frac{A_5 \rho}{g} \sqrt{2 g H} \sqrt{2 g H} = 2 A_5 \rho H$.

Pressure per square foot due to head $H = P_5 = \rho H$.

Therefore $2 P_5 A_5$ = momentum = impulse.

passes the overflow gap; hence, for steady motion, we have—

$$\frac{P_3}{\rho} + \frac{v_3^2}{2g} = \frac{P_4}{\rho} + \frac{v_4^2}{2g} \quad . \quad . \quad . \quad . \quad . \quad . \quad (39)$$

Now, $v_4$ can be neglected, being exceedingly small. This may be easily seen, thus: In an ordinary locomotive injector the diameter of the delivery cone orifice is about ¼in., while the orifice leading into the boiler is about 1½in. The ratio of the areas of these orifices is as 1 is to 36. Then, taking the velocity of the delivery jet to be about 144ft. per second roughly, the velocity of the water as it enters the boiler would be 4ft. per second, which can be neglected with respect to 144ft. per second.

Replacing

$$\frac{v_3^2}{2g} \text{ in equation (38)}$$

by

$$\frac{P_4 - P_3}{\rho} \text{ in equation (39),}$$

we find, after putting $A_5 = y\,A_2$ and $n\,P_2 = w\,P_1$, that

$$\frac{A_2}{A_3} = \frac{P_4\,(f + 2) - P_3\,(f + 1)}{w\,P_1 + 2\,y\,P_5} \quad . \quad . \quad . \quad . \quad (40)$$

This equation gives the ratio of the orifices for any injector. If the water is delivered into the boiler from which the motive power steam is taken, then $P_4 = P_1$, and

$$\frac{A_2}{A_3} = \frac{P_1\,(f + 2) - P_3\,(f + 1)}{w\,P_1 + 2\,y\,P_5} \quad . \quad . \quad . \quad . \quad (41)$$

If the overflow is open to the atmosphere, $P_3$ is the pressure of the atmosphere per square foot.

---

# CHAPTER VII.

## Weight of Water Delivered per Second.

The amount of water delivered to the boiler per second, say W = total water and steam delivered − steam delivered

$$= A_3\,\rho\,v_3 - z\,A_2\,\sqrt{P_1\,\rho_1} \quad . \quad . \quad . \quad . \quad . \quad . \quad (42)$$

The last term is obtained from equation (19) by putting $z$ for the coefficient given in the third column of Table I.

If we solve equation (38) for $v_3$, after first replacing $P_3$ by its equivalent, obtained from equation (39), remembering that $P_4 = P_1$, we have—

$$v_3 = \sqrt{\frac{2g}{\rho A_3 (1+f)}} \sqrt{(w P_1 + 2y P_5) A_2 - P_1 A_3} \quad . \quad (43)$$

Putting this in equation (42), we obtain—

$$W = \sqrt{\frac{2g\rho}{1+f}} \sqrt{(w P_1 + 2y P_5) A_2 A_3 - P_1 A_3^2} - z A_2 \sqrt{P_1 \rho_1}.$$

This may be written as—

$$W = A_3 \sqrt{\frac{2g\rho}{1+f}} \sqrt{(w P_1 + 2y P_5) \frac{A_2}{A_3} - P_1} - z A_2 \sqrt{P_1 \rho_1}.$$

Replace the ratio $\frac{A_2}{A_3}$ by its equivalent in equation (41), and we find that—

$$W = A_3 \sqrt{2g\rho (P_1 - P_3)} - z A_2 \sqrt{P_1 \rho_1} \quad . \quad . \quad . \quad . \quad (44)$$

We see at once that this is a maximum when $P_3$ is a minimum. For it to be possible to have a less pressure than that of the atmosphere, the connection with the atmosphere must be closed by a non-return valve, and the above result seems to indicate that the efficiency of the injector as a pump is increased thereby. The testimony of experiment also points in the same direction. We also obtain another advantage with a non-return valve, namely, that no air can be carried into the boiler with the jet of feed water.

In general, to deliver the feed water into the boiler at maximum temperature, the steam and water must be regulated to produce this result, but the cones are designed so as to give the maximum amount of feed water for a given quantity of steam, when the regulating device is not brought into action. Using the equation immediately preceding (44), we have the ratio—

$$\beta = \frac{\text{water}}{\text{steam}} = \frac{W}{z A_2 \sqrt{P_1 \rho_1}}$$

$$= \frac{A_3}{z A_2} \sqrt{\frac{2g\rho}{(1+f) P_1 \rho_1}} \sqrt{(w P_1 + 2y P_5) \frac{A_2}{A_3} - P_1} - 1 \quad . \quad (45)$$

This expression will be a maximum when

$$\frac{A_2 A_3 (w P_1 + 2 y P_5) - P_1 A_3^2}{A_2^2}$$

is a maximum. Taking the first differential coefficient of this expression with respect to $A_3$, and equating it to zero, assuming $A_2$ constant, we obtain the relation—

$$A_2 = \frac{2 P_1}{w P_1 + 2 y P_5} A_3 \quad . \quad . \quad . \quad . \quad . \quad . \quad (46)$$

Using this ratio in equation (46), we find that—

$$\beta \text{ max.} = \sqrt{\frac{2 g \rho}{\rho_1 (1 + f)}} \left[\frac{w P_1 + 2 y P_5}{2 z P_1}\right] - 1 \quad . \quad . \quad (47)$$

Also by combining equations (43) and (46), we see that when a maximum value of $\beta$ occurs—

$$v_3 = \sqrt{\frac{2 g P_1}{\rho (1 + f)}} \quad . \quad . \quad . \quad . \quad . \quad . \quad . \quad (48)$$

This result depends entirely upon the boiler pressure and the coefficient of resistance. From this equation we may obtain some limiting values of $f$ for maximum amount of feed water. When there is perfect condensation in the combining cone, the values of $f$ so found will probably be a maximum; but when perfect condensation does not occur—for instance, when the injector is delivering its minimum of feed water—then we should expect to find that the uncondensed steam would produce eddies in the water jet, and thereby increase the resistance. Referring to equation (39), we see that—

$$P_3 = P_1 - \rho \frac{v_3^2}{2 g} = P_1 - \frac{P_1}{1 + f} = P_1 \frac{f}{1 + f} \quad . \quad . \quad (49)$$

If $f$ is zero, then for a maximum value of $\beta$ there must be no pressure in the overflow chamber; and therefore there must be a non-return valve in the overflow pipe. The same may be inferred from an inspection of equation (44). In the same way, if W is a minimum, it appears that $P_3$ is a maximum, $A_2$ and $A_3$ being independent of $P_3$, as shown by equation (47). The ratio $A_2$ to $A_3$, given in this equation, is the one which must exist for a maximum amount of feed water; and therefore we may naturally conclude that all the steam is condensed, so that for average working we should expect to find the same resistance as in this particular case. Solve equation (47) for $f$, and we get—

$$f = \frac{P_3}{P_1 - P_3} \quad . \quad . \quad . \quad . \quad . \quad . \quad . \quad . \quad (50)$$

If $f$ were greater than here indicated, the injector would be doing more work than its maximum, or it would be supplying less than its maximum feed water, neither of which conditions can exist at the same time that equation (46) holds good. Hence we way naturally conclude that the value of $f$ here obtained will be a maximum when perfect condensation takes place. In many cases the overflow chambers are open to the atmosphere, and then the pressure there cannot be greater than 15 lb. per square inch. Putting this in the above, we obtain the following maximum values that can exist of $f$ for different boiler pressures :—

| Absolute boiler pressure per square inch. | Maximum value for $f$. |
|---|---|
| 25 | 1·5 |
| 40 | ·6 |
| 60 | ·33 |
| 75 | ·25 |
| 90 | ·20 |
| 105 | ·16 |
| 120 | ·14 |
| 150 | ·11 |
| 200 | ·08 |

In equation (38) we have put down the total resistance of friction, eddies, &c., to be $\frac{A_3 f \rho v_3{}^2}{2g}$.

Replace $f$ by its equivalent in (51), and substitute for the velocity $v_3$ from equation (39); we then get the maximum amount of resistance while the injector is delivering its maximum amount of feed water. It is $P_3 A_3$.

The annulus through which the water passes into the combining cone must be large enough to admit sufficient water per second to condense all the steam.

Weight of steam entering injector per second $= z\,A_2\,\sqrt{P_1\,\rho_1}$.
Heat in this steam above $t_3°$ Fah.

$$= (1{,}114 + \cdot 3\,t_1 - t_3)\,z\,A_2\,\sqrt{P_1\,\rho_1}\text{ thermal units.}$$

Weight of water required $= \beta\,z\,A_2\,\sqrt{P_1\,\rho_1}$

Heat required to raise the temperature of water from $t_5°$ to $t_3°$ Fah. $= (t_3 - t_5)\,\beta\,z\,A_2\,\sqrt{P_1\,\rho_1}$ thermal units.

Then, as the heat gained by the feed water must be the same as that lost by the steam, we have

$$\beta = \frac{1{,}114 + \cdot 3t_1 - t_3}{t_3 - t_5}$$

a relation found on a previous occasion.

Total amount of feed water flowing through the water annulus per second under pressure $P_5$

$$= y\,A_2\,\sqrt{2\,g\,\rho\,P_5}$$

Therefore total heat supplied to feed water per second

$$= (t_3 - t_5)\,y\,A_2\,\sqrt{2\,g\,\rho\,P_5}\text{ thermal units.}$$

Hence

$$y = \frac{(1{,}114 + \cdot 3\,t_1 - t_3)\,z\,\sqrt{P_1\,\rho_1}}{(t_3 - t_5)\,\sqrt{2\,g\,\rho\,P_5}} \quad . \quad . \quad . \quad . \quad . \quad (51)$$

$$= \beta\,z\,\sqrt{\frac{P_1\,\rho_1}{2\,g\,\rho\,P_5}} \quad . \quad . \quad . \quad . \quad . \quad . \quad . \quad . \quad (52)$$

The value of $y$, thus determined, pre-supposes perfect condensation. As the action does not generally approach so near the ideal as this, we must give $y$ a larger value. It ranges between 2 and 3 in practice, with fixed nozzles and high pressures.

In the above equation (52), if we assume that the pressure of the water is one-third that of the atmosphere (*i.e.*, equivalent to a lift of 20 ft.), $y$ then becomes nearly 2.

After measuring a number of injectors, the author finds that $A_5 = 2{\cdot}75\,A_2$ on the average. A reason for this apparent discrepancy is to be found, the author thinks, in the fact that we have assumed the condensation perfect, and ignored the pressure of the water vapour in the combining cone. With large quantities of feed water, or low temperatures of delivery, this will be inappreciable; but with high temperature delivery it may approach the atmospheric pressure. In the latter case the feed water will approach the combining cone under a less pressure than before, and therefore with

TABLE V.

Maximum weight of feed water delivered per hour in pounds when the overflow is open. The steam for the injector is taken from the same boiler into which the feed water is delivered.

| Size of the combining cone orifice in millimetres. | Size of the combining cone orifice in inches. | Pressure of steam in pounds per square inch. | | | | | | | | | | | | |
|---|---|---|---|---|---|---|---|---|---|---|---|---|---|---|
| | | 20 | 30 | 40 | 50 | 60 | 70 | 80 | 90 | 100 | 110 | 120 | 130 | 140 |
| | | Delivery of feed water in pounds per hour. | | | | | | | | | | | | |
| 2 | ·0787 | 350 | 430 | 500 | 560 | 610 | 660 | 710 | 750 | 800 | 830 | 870 | 910 | 930 |
| 3 | ·1181 | 800 | 970 | 1120 | 1260 | 1380 | 1500 | 1600 | 1700 | 1780 | 1870 | 1960 | 2030 | 2110 |
| 4 | ·1574 | 1420 | 1730 | 2010 | 2250 | 2460 | 2660 | 2850 | 3010 | 3170 | 3330 | 3480 | 3620 | 3760 |
| 5 | ·1968 | 2220 | 2720 | 3130 | 3510 | 3850 | 4160 | 4450 | 4710 | 4960 | 5220 | 5450 | 5660 | 5870 |
| 6 | ·2362 | 3200 | 3920 | 4520 | 5060 | 5550 | 5980 | 6400 | 6780 | 7150 | 7500 | 7830 | 8160 | 8460 |
| 7 | ·2755 | 4350 | 5330 | 6160 | 6880 | 7550 | 8150 | 8710 | 9230 | 9730 | 10210 | 10670 | 11100 | 11520 |
| 8 | ·3148 | 5680 | 6960 | 8030 | 8980 | 9850 | 10620 | 11370 | 12060 | 12720 | 13350 | 13930 | 14500 | 15050 |
| 9 | ·3543 | 7200 | 8820 | 10170 | 11380 | 12470 | 13470 | 14400 | 15270 | 16100 | 16880 | 17630 | 18350 | 19050 |
| 10 | ·3936 | 8880 | 10880 | 12560 | 14050 | 15400 | 16630 | 17770 | 18870 | 19870 | 20860 | 21770 | 22660 | 23520 |
| 11 | ·4330 | 10750 | 13170 | 15200 | 17000 | 18630 | 20120 | 21500 | 22810 | 24050 | 25220 | 26330 | 27420 | 28460 |
| 12 | ·4724 | 12800 | 15670 | 18100 | 20230 | 22170 | 23950 | 25600 | 27150 | 28610 | 30020 | 31360 | 32630 | 33870 |
| 13 | ·5127 | 15010 | 18400 | 21230 | 23520 | 26020 | 28100 | 30050 | 31870 | 33580 | 35230 | 36800 | 38300 | 39750 |
| 14 | ·5510 | 17410 | 21330 | 24620 | 27530 | 30180 | 32610 | 34850 | 36960 | 38950 | 40870 | 42670 | 44410 | 46100 |
| 15 | ·5904 | 19980 | 24500 | 28270 | 31610 | 34650 | 37420 | 40000 | 42420 | 44710 | 46910 | 49000 | 50980 | 52920 |
| 16 | ·6296 | 22750 | 27870 | 32170 | 35970 | 39420 | 42570 | 45510 | 48270 | 50870 | 53370 | 55750 | 58010 | 60220 |
| 17 | ·6691 | 25670 | 31460 | 36310 | 40610 | 44500 | 48060 | 51380 | 54500 | 57430 | 60260 | 62910 | 65480 | 67980 |
| 18 | ·7086 | 28780 | 35270 | 40710 | 45520 | 49900 | 53880 | 57600 | 61100 | 64380 | 67560 | 70550 | 73420 | 76330 |
| 19 | ·7479 | 32070 | 39300 | 45360 | 50720 | 55600 | 60030 | 64180 | 68070 | 71750 | 75270 | 78610 | 81800 | 84920 |
| 20 | ·7872 | 35530 | 43550 | 50260 | 56210 | 61600 | 66520 | 71100 | 75420 | 79500 | 83410 | 87100 | 90610 | 94100 |
| 21 | ·8266 | 39180 | 48030 | 55410 | 61960 | 67910 | 73330 | 78100 | 83860 | 87600 | 91670 | 95410 | 98470 | 101860 |
| 22 | ·8660 | 43000 | 52680 | 60810 | 68010 | 74520 | 80480 | 85710 | 92050 | 96150 | 100610 | 104710 | 108070 | 111800 |
| 23 | 9054 | 46830 | 57420 | 66380 | 74120 | 81460 | 87570 | 93670 | 99380 | 105080 | 109970 | 114860 | 118120 | 122180 |
| 24 | ·9448 | 51000 | 62520 | 72280 | 80710 | 88700 | 95350 | 102000 | 108210 | 114420 | 119750 | 125060 | 128600 | 133050 |
| 25 | ·9841 | 55330 | 67850 | 78430 | 87580 | 96250 | 103460 | 110680 | 117420 | 124150 | 129930 | 135710 | 139560 | 144370 |

**Note.**—The nominal horse power multiplied by 100 gives approximately the number of pounds of feed water per hour.

The figures in Table V can be obtained very nearly by the following equation

W = lbs of water per hour d/d into boiler

P = Abs. pressure per sq. in. in boiler.

(1) d = dia of throat in m/m

(2) D = dia of throat in inches.

(3) $D_1$* = " " " " hundredths of an inch

Throat = d/y throat

$$W = d^2 \times \sqrt{P} \times 20 \quad \ldots (1)$$

$$W = D^2 \times \sqrt{P} \times 13000 \quad \ldots (2)$$

m/m = ·03937 inch

$$W = D_1^2 \times \sqrt[2]{P} \times 1{\cdot}30 \quad \ldots (3)$$

Distance from point of Steam Cone to point of Combining cone is not stated in this book. Eaton gives 5 dias of steam cone. One of my small flap nozzle injectors had 6·7 dias and worked well. The Exhaust Injector Co's No 3 Injector has 7.2 dias of the mouth of the steam Cone which is ·25" dia as compared with ·183" dia higher up in the jet. See sketch on loose sheet.



(over)

Nov 2/10

One of my Injectors with a throat 2·1 100$^{ths}$ dia put in 49 lbs of water per hour (at the rate of) at 60 lbs gauge pressure.

By formula (3) $2{\cdot}1^2 \times \sqrt{75} \times 1{\cdot}30 = 49{\cdot}63$ lbs which very closely agrees with the actual performance, even in so small an injector as this.

less velocity; hence the water annulus must be correspondingly increased. If the entrance to the combining cone is not well rounded, and gently tapered, a certain amount of contraction of the water jet will take place as it enters, and thus the net sectional area of the jet will be less than the orifice, and therefore under these circumstances a larger orifice will be required. More will be said about the pressure in the combining cone when we come to treat of the exhaust injector.

So far the author has not heard of any experiments made to determine this pressure, though he hopes to carry out some himself in the near future. It is difficult to treat the action of the water and steam in the combining cone mathematically, as we are to a certain extent ignorant of data to work upon. We can hardly assume that the motion is steady, as one would be led to expect something much different.

The ratio of $A_2$ to $A_3$ depends upon the value of $y$ in equations (40), (41), and (46). It would be preferable to obtain this ratio independent of $y$ altogether. We may accomplish this as follows :—

It has been shown that

$$(1 + \beta) = \frac{1{,}114 + {\cdot}3\,t_1 - t_3}{t_3 - t_5} + 1 = \frac{1{,}114 + {\cdot}3\,t_1 - t_5}{t_3 - t_5}.$$

Also, by combining equations (43), (44), and (48), we find that

$$(1 + \beta) = \frac{A_3}{z\,A_2}\sqrt{\frac{2\,g\,\rho}{\rho_1\,(1 + f)}} \text{ or } \frac{A_3}{z\,A_2}\sqrt{\frac{2\,g\,\rho\,(P_1 - P_3)}{P_1\,\rho_1}} \quad . \; . \; (53)$$

Therefore

$$\frac{A_2}{A_3} = \frac{1}{z(1+\beta)}\sqrt{\frac{2\,g\,\rho}{\rho_1\,(1+f)}} = \frac{1}{z}\left[\frac{t_3 - t_5}{1{,}114 + {\cdot}3t_1 - t_5}\right]\sqrt{\frac{2\,g\,\rho}{\rho_1\,(1+f)}} \quad (54)$$

From this we gather that with an exhaust injector the steam cone must be exceptionally large—in fact, it must be inversely proportional to the square root of the density of steam at atmospheric pressure. The same may also be inferred from an inspection of equation (38).

In any injector, whether live steam is used or not as the motive power, the steam orifice will vary inversely as the square root of the density of the boiler steam; that is, approximately inversely as the square root of the boiler pressure. At the same time, equations (49) and (50) indicate that $y$ varies directly as the pressure roughly. Therefore the ratio of the feed water orifice to steam orifice must vary directly as the boiler pressure, and directly as the quantity $\beta$. This shows that for varying steam pressures it is necessary for maximum efficiency to regulate the steam and water orifices relatively to each other. This may be done by either keeping the steam constant and altering the water supply, or by keeping the water constant and altering the steam, or the both may be altered together. This latter method, if it can be done easily and by simple means, seems to commend itself. For the sake of convenience, we may collect the results we have just obtained. They are for a live steam injector :—

(1). Water and steam together delivered to the boiler per second

$$= A_3 \sqrt{2 g \rho (P_1 - P_3)} \text{ lb.}$$

2). Feed water delivered to the boiler per second

$$= W = A_3 \sqrt{2 g \rho (P_1 - P_3)} - z A_2 \sqrt{P_1 \rho_1} \text{ lb.}$$

(3). The ratio of the steam to delivery orifice

$$= \frac{A_2}{A_3} = \frac{1}{z(1+\beta)} \sqrt{\frac{2 g \rho}{\rho_1 (1+f)}} = \frac{1}{z} \left[ \frac{t_3 - t_5}{1114 + \cdot 3 t_1 - t_5} \right] \sqrt{\frac{2 g \rho}{\rho_1 (1+f)}}$$

The author has found, by measuring a number of injectors, that the combining cone orifice is slightly larger than the diverging cone orifice when there is a large overflow gap, such as in non-lifting injectors with fixed nozzles. This the author accounts for as follows : The jet as it issues from the combining cone is a converging jet, and it has been diminishing in diameter as it proceeded down the cone; hence it will go on diminishing to a certain extent after leaving the orifice, similar to a jet issuing from a sharp-edged orifice, though less in degree; therefore when it arrives at the diverging cone orifice it will be smaller in cross section than when it left the combining cone. On the other hand, the author has found that with injectors which have a small overflow gap, such as the flap nozzle injector, the two cones are identically the same in size at the orifice. The above results will hold for the exhaust injector also, if we write $P_a$ and $\rho_a$ for $P_1$ and $\rho_1$ where they are associated with $A_2$.

As an example of the use of the above, we propose to find the quantity of water and steam delivered to a locomotive boiler when the absolute pressure is 140 lb. per square inch, the diameter of the combination cone orifice 8 millimetres, the temperature of the feed water 60 deg. Fah., the temperature of the delivery 190 deg. Fah. The overflow supposed open to the atmosphere.

The weight of water and steam delivered per second

$$= A_3 \sqrt{2 g \rho (P_1 - P_3)}$$

A millimetre is ·03937 in.; hence the diameter of the orifice

$$A_3 = 8 \times ·03937 = ·315 \text{ in.}$$

$$A_3 = \frac{\pi}{4} \times ·315^2 = ·0782 \text{ square inch.}$$

$$= \frac{·0782}{144} \text{ square feet.}$$

Then

$$A_3 \sqrt{2 g \rho (P_1 - P_3)} = \frac{·0782}{144} \sqrt{64·4 \times 62·5 \times 144 (140 - 15)}$$

$$= 4·615 \text{ lb. per second.}$$

$$= 16614 \text{ lb. per hour.}$$

The ratio of feed water to steam depends upon the temperatures of supply and delivery, and they are connected by the equation

$$\beta = \frac{1114 + ·3\, t_1 - t_3}{t_3 - t_5}$$

$$= \frac{1114 + (·3 \times 352) - 190}{190 - 60} = 7·9$$

Therefore the weight of water and steam is to the weight of steam as $(1 + 7·9) : 1$.

Hence weight of steam supplied per hour

$$= \frac{\text{delivery}}{1 + \beta} = \frac{16614}{8·9} = 1875 \text{ lb.}$$

Net amount of feed water delivered per hour

$$= 16614 - 1875 = 14739 \text{ lb.}$$

If it is required to obtain the maximum amount of feed water that will enter the boiler per hour, we can refer to Table IV., and we there see that each pound of steam will deliver 9·6 lb. of feed water. We then have the weight of steam supplied per hour

$$= \frac{\text{delivery}}{1 + \beta} = \frac{16614}{10·6} = 1570 \text{ lb.}$$

Net amount of feed water per hour

$$= 16614 - 1570 = 15044 \text{ lb.}$$

On referring to the maximum quantity of water which several manufacturers guarantee that their No. 8 injectors will supply, the author finds a peculiarly striking confirmation of the above result. Three numbers taken from some of the best manufacturers are 15,000, 15,000, and 15,050. Following on with the above problem, we have the weight of steam used per second $= \frac{18750}{3600}$ lb.

$$= z A_2 \sqrt{P_1 \rho_1} = 3{\cdot}4 \times A_2 \sqrt{144 \times 140 \times {\cdot}315}$$

from which we obtain $A_2 = {\cdot}276$ square inches, after reducing the square feet to inches. The diameter of $A_2$ will then be ·593 in.

The ratio $A_2$ to $A_3$ is 3·5, and the ratio $d_2$ to $d_3$ is 1·88.

If we assume the area $A_5$ to be 2·75 $A_2$, then

$$A_5 = 1{\cdot}62 \text{ square inch.}$$

It will be noticed that the expression for the weight of water and steam delivered to the boiler per second depends upon the pressure in the space into which the water is delivered, and also upon the pressure in the combining cone. Upon reflection it will be seen, apart from the analysis just gone through, that such must be the case, for the delivery jet must have a large enough velocity to force its way into the boiler. The velocity of a jet of water issuing from the boiler through a pipe, the pressure at the end of which is $P_3$, is given by the equation

$$\frac{P_4}{\rho} = \frac{P_3}{\rho} + \frac{v_3^2}{2g}$$

$$\text{hence } v_3 = \sqrt{\frac{2g(P_1 - P_3)}{\rho}}$$

A jet of water, to enter the boiler at all, must have a velocity not less than $v_3$. With this velocity the weight of water which enters the boiler per second is $A_3 \rho v_3$

$$= A_3 \sqrt{2 g \rho (P_1 - P_3)} \text{ lb.}$$

The different values for the weight of feed water in pounds delivered per hour are given in the appended table.

As each nominal horse power of a boiler requires about 100 lb. of feed water per hour, we may multiply the nominal horse power of a boiler by 100 to get the amount of feed water required by it per hour, and then see in the table what size injector will give the requisite amount of feed with a given steam pressure.

The result of Table V. may be shown perhaps with great clearness by plotting a curve for each different size injector, having as ordinates the nominal horse power of the boiler to be fed, and the absolute boiler pressure in pounds per square inch as abscissæ. A set of 25 curves will be found in figures 4 and 5. The two figures should be really in one, but the great disparity in the extreme cases necessitated the first fourteen sizes being placed together, while the eleven larger sizes have been plotted to a smaller scale. It will be noticed that the ordinates of nominal horse power have been reckoned from a horizontal line at the top of the diagram, instead of at the base, as is customary. The reader will find little difficulty in using the curves. Let it be required to find the size of the injector which is to supply boilers of 200 collective horse power, when the pressure in a boiler is 100 lb. per square inch. In fig. 4 look out the horizontal line corresponding to 200 horse power. Then find the intersection of this line, and the ordinate corresponding to 100 lb. pressure. It will be noticed that this happens nearly on the curve marked 10, on the right-hand side of figure, which shows that a No. 10 injector will be required to feed the boilers, or that the diameter of the delivery cone at its smallest section is 10 millimetres. Take another case. A 50 horse power boiler is to be supplied with feed water, when the boiler pressure is 130 lb. per square inch. The abscissa, 50 horse power, intersects the ordinate 130 lb. between the curves 4 and 5; hence the required work is too much for a No. 4 injector, and therefore a No. 5 must be used. These examples are enough to show how easily an injector may be selected by consulting the curves or the table.

Of course, the quantities given in either the table or the curves correspond to maximum delivery per pound of steam, and hence the water must be supplied to the injector *cold*—*i.e.*, not above 60 deg. Fah. With fixed nozzle injectors this maximum delivery only happens at a particular boiler pressure, namely, that for which the injector was designed; but with injectors that can be regulated by a relative movement of the cones, the maximum values given above can be obtained at all pressures. A curious point presents itself here. We will suppose an injector designed for maximum

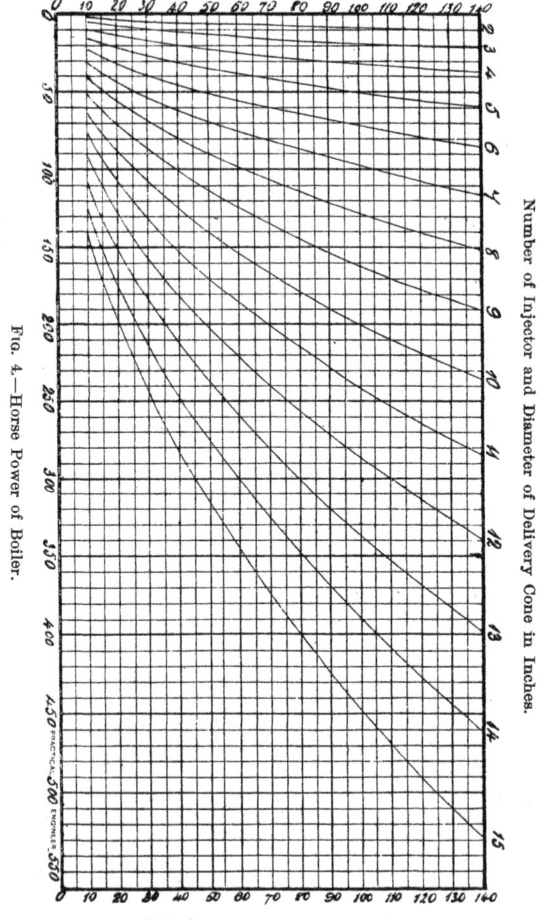


Fig. 4.—Horse Power of Boiler.

quantity of feed water per pound of steam. The delivery

cone must, of necessity, be large enough to permit of the full quantity of water and condensed steam passing through it. With maximum feed water we have a minimum velocity of delivery, both conditions together requiring the delivery cone to be of maximum diameter. Now, suppose that the injector is made (by regulation) to deliver the minimum

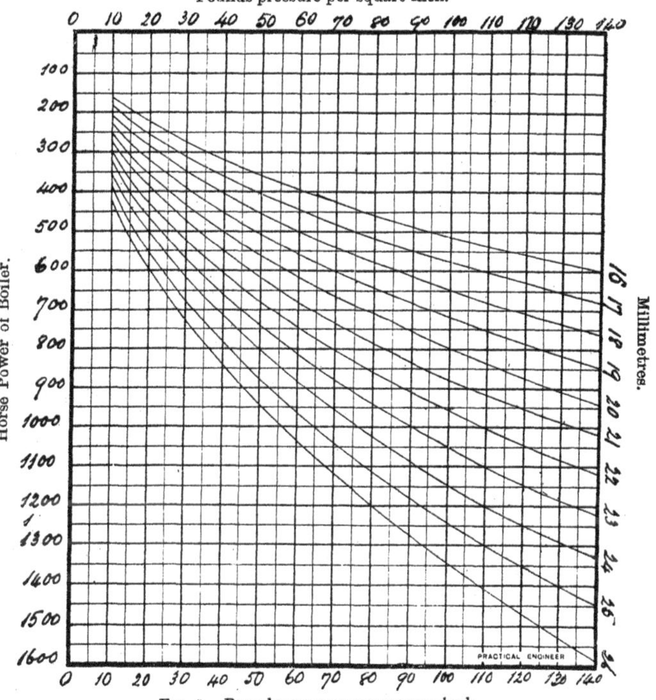


FIG. 5.—Pounds pressure per square inch.

quantity of feed per pound of steam. The velocity of the delivery jet will be increased, while the quantity of water passing the smallest part of delivery cone will be less than before ; therefore the delivery cone is not required to be so large as in the previous case. If it remains constant, as it does always in practice, then the only result which can follow is that it is only partially filled with water. This conclusion appears antagonistic to the general laws of hydraulics, where the pipe must be full before a reduction of velocity is accompanied by a corresponding increase in pressure. A possible reason may be found in the assumption that the steam is not wholly condensed in the combining cone, which, together with the vapour given off from the water at the temperature of delivery, produces a pressure in the delivery cone surrounding the jet of water sufficient to preserve steady motion. The result of experience also seems to point towards the same result. On reflection, after

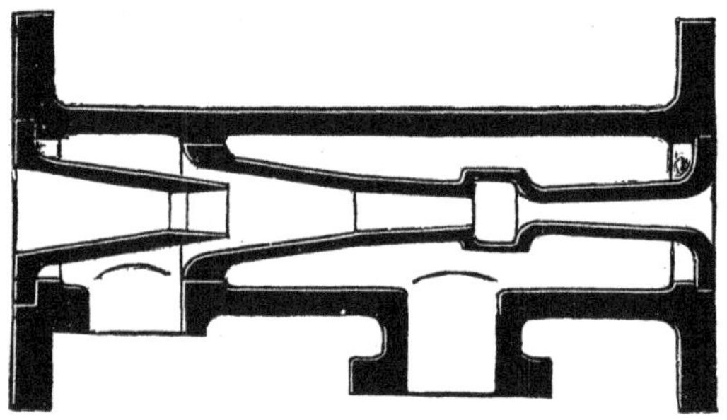

Fig. 6.

considering the above conclusions, it would appear that a regulation of the steam and water cones should be accompanied by a corresponding regulation of the delivery cone. The same conclusions would be gathered from a perusal of the previous mathematical passages. Such regulation has not yet been attempted in practice, so far as the author is aware.

---

## CHAPTER VIII.

It is proposed, in what immediately follows, to give some representative examples of the different forms of injectors now in use, and to briefly point out the chief features of each particular form of apparatus. In such a description, it is needless to say that there may be points which have escaped the author's notice, and there must be—amongst the large number of manufacturers in these times of keen competition—many omissions of examples from the total list of injector makers. The object of these notes is to supply a concise and as complete account as possible of the working of the injector, without in any way being trammeled with a choice of examples or influenced by the

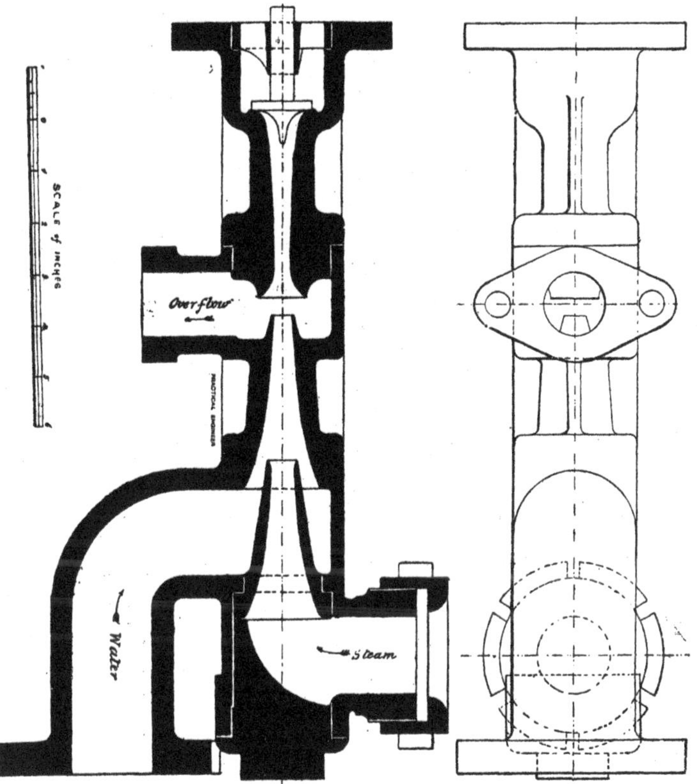


FIG. 7.—Standard Injector, Great Western Railway.

highly eulogistic paragraphs generally to be found in sale catalogues. At the same time, the author again wishes to express his deep sense of indebtedness to those who have so kindly lent their blocks, and forwarded other illustrations for the purpose of assisting him in illustrating his descriptions.

The afore-mentioned keen competition, and the excessive craze for cheapness which is every day evinced by private firms and the directors of public companies, have led to the publication of much matter which—to use a mild term—we may describe as considerably strained, in connection with injectors and their performances; and it is not an uncommon event for an injector to give a deal of trouble, through the bad design and lack of adjustment of unscrupulous makers.

The universal application of the injector, as a boiler-feeding apparatus, to the locomotive, will, the author thinks, excuse him in dwelling at some length upon the development of that particular form, and for the many sketches of it which may be given.

---

## CHAPTER IX.

### NON-LIFTING INJECTORS.

IN this class of apparatus, it is absolutely essential—as the name implies—that the cold feed water source should be at least as high as the injector itself, and better results are obtained when it is a few feet above. A typical example is given in fig. 6, which is of the simplest form possible, being merely the outside casing (generally of cast iron), with the steam cone, combining cone, and diverging cone fitted into sockets without any of the usual accessories. The combining and diverging cones are cast in one, the overflow gap dividing them. They are secured by the delivery-pipe flange, while the steam cone is secured by the steam-pipe flange. It will be noticed that the top of the combining cone merely slides into the casing, because it is not necessary to the efficient working of the injector that it should be absolutely tight under pressure. At the greatest, there can only be the pressure of the atmosphere there, and should there be a slight leakage, it is harmless, and would soon be stopped by sediment. For the accompanying sketch the author is indebted to the Exhaust Injector Company (successors to Messrs. Sharpe, Stuart, and Co., who first introduced them), but probably every maker throughout the kingdom produces a similar article. In one locomotive injector, the author has found the following diameters of the cones at their smallest sections:

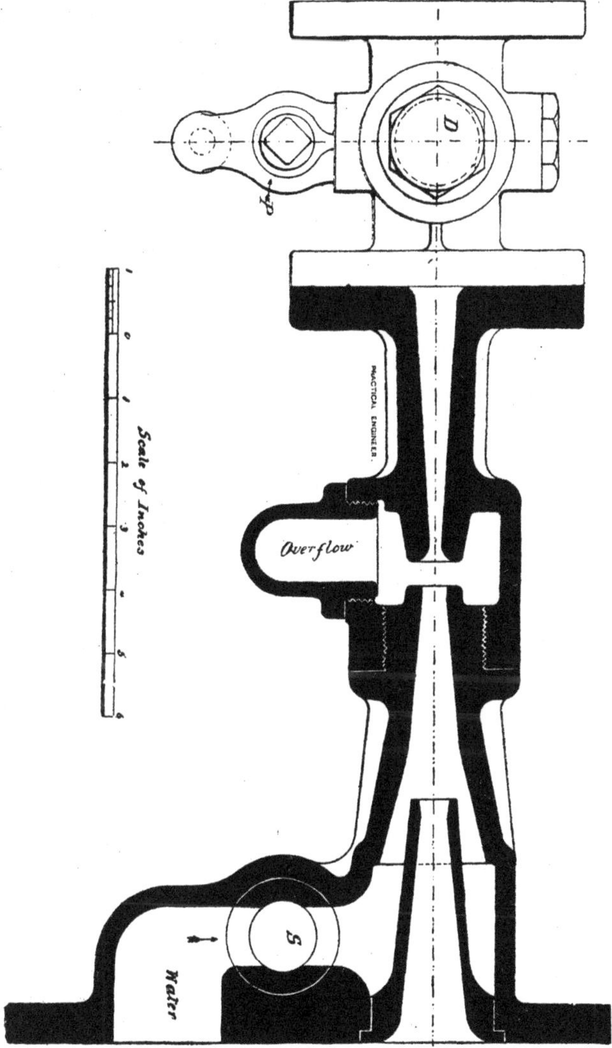


FIG. 8.—Section of Standard Goods Engine Injector, Great Eastern Railway.

Steam cone, 12·8 millimetres; combining cone, 10·4 millimetres; and diverging cone, 8 millimetres. The overflow gap in this case was large, and if a generator of the combining cone were produced, it would touch the diverging cone at its minimum section. The cones in all injectors are always made of brass, on account of the corrosion and rust that would take place with iron.

There must be stop valves on the steam and water pipes, as there are none attached to the injector itself; and these valves are also required for regulating the supply of water and steam, so that there may be no overflow during the time the injector is working. These non-lifting injectors are sometimes troublesome in starting, unless they are nicely designed, and working with the particular steam pressure for which they were designed. They are also liable to be stopped by the jolting of a locomotive as it passes over crossing points, when it has to be re-started by hand in the usual manner, namely, by turning the water full on, and then opening the steam valve a little until the water is seen to issue from the overflow with increased velocity. This is an indication that the injector has "caught on," as the starting is termed, and then the steam valve should be opened full. If any water then comes from the overflow pipe, the water valve should be partially closed until the overflow ceases.

Mr. A. L. Sacré, M.I.C.E., of the firm of Gresham, Craven, and Co., who have had a very extensive experience in designing injectors, informs the author that a well-constructed injector, designed for the purpose of dealing with water at a high temperature, could take the feed water at the following maximum temperatures:—

| | | | | | |
|---|---|---|---|---|---|
| 145 deg. | Fah., | with | a steam | pressure of | 30 lb. |
| 130 | „ | „ | „ | „ | 75 „ |
| 120 | „ | „ | „ | „ | 100 „ |
| 115 | „ | „ | „ | „ | 120 „ |
| 100 | „ | „ | „ | „ | 150 „ |

There is one serious drawback to the particular form of injector shown in fig. 6, when used in a district in which the feed water carries a lot of lime and magnesia in solution, and that is, the cones cannot be removed from the casing without breaking the joints, and perhaps removing the injector to the workshop to remove the scale.

It is as well, with all injectors, to use a copper pipe for the overflow, otherwise it should be large in comparison with the steam pipe, because the rust inside the pipe

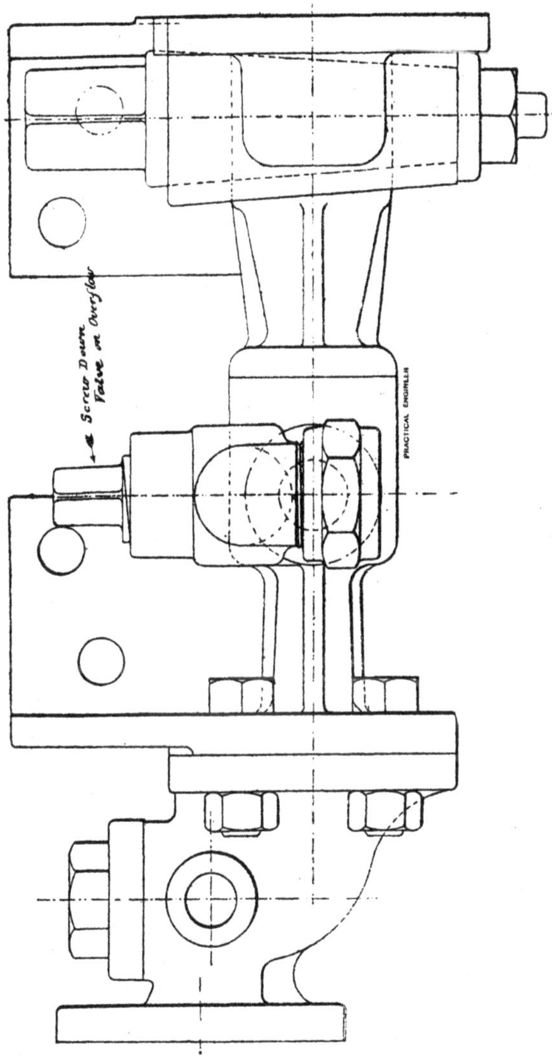


FIG. 8.—Elevation of Standard Goods Engine Injector, Great Eastern Railway.

assists the sediment in depositing, and where bad water is used the pipe is liable to become quite or nearly closed, thereby rendering the injector useless until it is removed.

A case of this sort came under the author's notice some years ago. A tank locomotive, having one of these non-lifting injectors placed by the side of the firebox, under the foot plate, was working for a time in a very bad lias limestone district. Sometime after the engine had left the shops, the injector gave a good deal of trouble, and soon stopped altogether. After many joints were broken and re-made, the overflow pipe (which had a bend near the flange) was removed, and found to be almost completely choked with a deposit at the bend. After its removal the injector caused no further trouble. The author has been enabled, through the kindness of Mr. W. Dean, M.I.C.E., locomotive and carriage superintendent, to give in fig. 7 a reproduction of the standard injector used on the Great Western Railway for both goods and passenger engines. Two views are given, an elevation and sectional plan. The outside casing in the previous figure has been dispensed with, and the necks of the cones strengthened by ribs. The combining cone, the overflow pipe flange, the water and steam-pipe flange are contained in one casting, the steam cone being kept in position by the screw plug above it. The object of the peculiar shape of the plug is to provide a continuous contour to the steam channel. The back-pressure valve fits down upon the end of the diverging cone, and has a tapering projection on the side next to the cone. This projection guides the water from its previous direction down the cone into the chamber beyond the valve, without producing eddies, as otherwise would happen and therefore a loss of head, if it were not there. Mr. Dean has apparently designed the cones so that the velocity of the fluid shall be accelerated uniformly as it passes down them. This would seem to be as reasonable as any other method, and perhaps better.

In fig. 8 we have a side elevation and sectional plan of the injector which Mr. Jas. Holden, M.I.C.E., uses upon the goods engines of the Great Eastern Railway. This injector is very similar to the one just illustrated, except that it is made in two parts, instead of three, and contains its own water-regulating valve. Two of these injectors are attached to each engine, one of them only having a valve on the overflow orifice (fig. 9), with a screw-down arrangement above it. The primary object of this attachment is to close the overflow aperture, and admit steam to the water in the tender

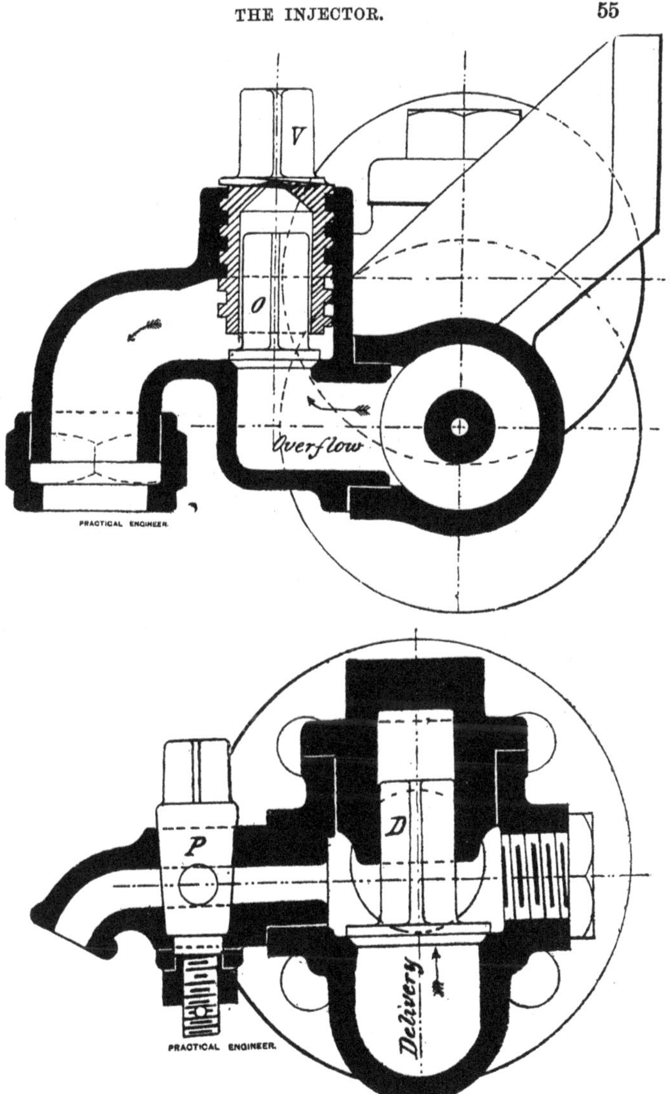


FIG. 9.—Overflow Valve and Delivery Valve.

or tanks when it is blowing off at the safety valve; which heat will be stored up, minus the little radiation that takes place, until required to be fed into the boiler. This is a useful attachment, to goods engines especially, as they often have to wait in sidings until passenger trains have passed before they can proceed on their journey, during which time the safety valve will be probably blowing off considerably. It will be noticed that the valve is separate from the screw-down arrangement. This prevents any air being drawn into the boiler by the injector, and somewhat increases the delivery.

A non-return delivery valve D is placed in a little casting bolted to the diverging cone, having a pet cock P. The chamber and valve are shown in section (fig. 9). The injector is supported in a horizontal position by two brackets, shown in the elevation. Both the screw-down valve on the overflow and the water regulator S are connected by rods to handles on the foot-plate, so that they can be easily operated by the fireman or driver.

Two longitudinal sections of the standard injector used on the London and North-Western Railway by Mr. F. W. Webb, M.I.C.E., are shown in fig. 10. The injector is fixed below the foot-plate, with the diverging cone pointing vertically upwards, and connected by a copper pipe with the clack box, which is immediately above the injector, and fixed to the back outside firebox plate, so as to be within easy reach of the fireman. The characteristic boldness of design exhibited by most things that emanate from the brain of Mr. Webb is not wanting even in this injector. It is the almost universal rule in practice to keep the diverging cone stationary, and regulate the flow of steam with a taper plug in the steam cone, or the flow of water by an axial movement of the steam cone; but in this particular instance the combining and diverging cones are cast in one piece, and made to slide axially by means of the rod and handwheel W, thereby regulating the supply of water. At the same time the rod operates the sliding cones through the medium of the overflow valve; and by a horizontal motion of the handle H, fixed on the rod, the overflow orifice is closed to the atmosphere, and steam may be blown through the injector into the tender. The outside casing of the overflow valve is attached to the sliding cones by means of a nut, as shown in the figure. The injector is bolted to the outside firebox plate. A pet cock P is situated a little distance below the clack box, to admit of damping the coal or washing down

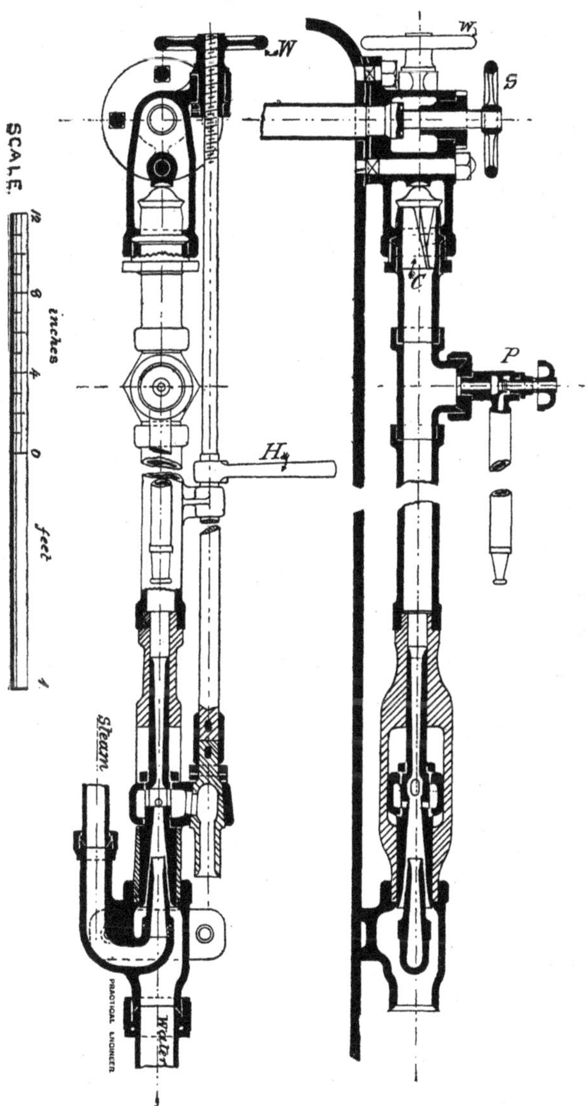


Fig. 10.—Standard Injector, L. and N.W. Railway, F. W. Webb's Patent

the foot-plate, when necessary. The clack box contains, besides the ordinary clack valve, and independent of it, a screw-down valve S, so that the clack can be examined when under steam. The clack valve C has spiral wings, so that it is continually being turned round by the incoming water every time it is opened, and thus wears the seating uniformly all round. Mr. Webb has informed the author that they now have a new re-starting automatic injector, which they are fixing in place of the above, leaving the clack box, stand pipe, and connections just as they are at present.

It may be asked, and with good reason, too, why will not a non-lifting injector lift its own feed water, instead of having it fed to it from a higher level? This is a point which seems to have been missed by most writers upon the subject, though why it has been it is difficult to say.

The capacity of an injector to lift its own feed water depends entirely upon the capacity of the steam jet to exhaust the water pipe of air; and therefore the lifting injector must be so designed as to perform the part of an *ejector* until the feed water has reached the combining cone, and then it must be capable of fulfilling the functions of an injector, to force the water into the boiler. The two separate operations depend upon entirely different principles. The action of the injector we have already analysed, and we will now proceed to show the necessary provisions to be made so that the injector may be lifting, instead of non-lifting. If we turn back to the investigation of flow of a fluid through a closed channel (when the fluid completely fills the channel), we see from equation (3) that when water is the fluid, the pressure per square foot divided by the density + the square of the velocity divided by $2.g$ equals a constant; or, in other words, the pressure will decrease as the velocity increases, and *vice versa*. The same thing happens in a greater or less degree with any fluid, whether gaseous or liquid, as is indicated in equation (14). In fig. 11 we have a diagrammatic view of the steam and combining cones of a non-lifting injector. The previous analysis tells us that the steam cone orifice must be larger than the combining cone orifice—that is, that A is greater than B. But if the injector is to lift its feed water, it must exhaust the water pipe and chamber at beginning of combining cone of air—that is, the absolute pressure in the water pipe and the space surrounding A must be less than that at B, which is 15 lb. per square inch, approximately. But by the laws of the flow of fluids, before mentioned, the pressure in the space surrounding A will be greater than that at B, because the velocity is less than at B, on account of the larger diameter; hence, instead of producing a vacuum in the water pipe with the arrangement in fig. 11, we have an increase in the pressure, and therefore it is impossible to produce a vacuum there with this arrangement, so as to lift the feed water. Now, instead of the arrangement shown (fig. 11), let us have that shown (fig. 12) in which the right-hand cone *diverges*, instead of converges, from the orifice A, and in which the minimum diameter is greater than the diameter of the orifice A. The pressure at

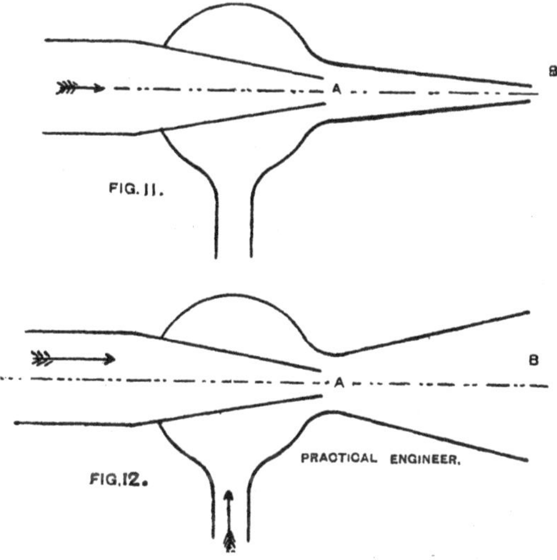


B is that of the atmosphere, and as the orifice A is approached the diameter diminishes, therefore the velocity increases, which must be accompanied by a decrease in pressure [equations (3) and (14)]. From this it is easily seen that to allow of an injector lifting its feed water it must have, between the steam cone orifice and the combining cone orifice, an opening considerably larger than the steam cone orifice itself; otherwise the water pipe will not be partially exhausted, and no water will reach the injector. There is another way in which this can be accomplished; or perhaps it would be more correct to say there is another direction in the same way along which we may proceed to this end.

We have seen that the steam orifice must be considerably less than the orifice by which the steam escapes into the atmosphere while the water pipe is being exhausted. In the previous method we maintained the steam cone orifice constant, and enlarged the combining cone orifice. We may still keep the same ratio of steam orifice to escape orifice, if the overflow orifice is maintained constant, and the steam cone diminished by sliding axially into it a solid cone, thus partly filling up the orifice. How these two methods are carried out in different forms of injectors the author will now proceed to show.

---

## CHAPTER X.

### Lifting, Automatic, and Re-starting Injectors.

The lifting injector, of which a longitudinal section is given in fig. 13, is almost identical with that first introduced by the original inventor, M. Gifford, something over thirty years ago. The body of the instrument is made in three separate castings, all of which, in this particular instance, are screwed together. The topmost of these contains the flanges of the water and steam inlets, and terminates in a gland and stuffing box, through which slides axially the hollow rod, on the end of which is screwed the steam cone. This hollow rod has two or more openings opposite the steam inlet in the outer casing to allow of the steam passing from the steam pipe to the steam cone. The orifice of the steam cone can be varied at pleasure by the tapered spindle running down through its centre, having a square screw thread turned upon it, about half way along its length. Above the thread we have a collar and a small stuffing box and gland to secure the spindle steam-tight. A small hand wheel placed on the upper extremity of the spindle enables the steam cone orifice to be regulated in area as nicely as desired, while the regulation is unaffected by any axial motion of the hollow spindle. The combining and diverging cones are similar to those in the previously described non-lifting types, though they are of extra length. A non-return or back-pressure valve has its seating on the end of the diverging cone, and is of the form generally used for that purpose, having a projection turned, the part facing the cone, to guide the stream lines without producing eddies and thereby causing a resistance to the flow of water from the injector. The narrow portions of the cones are strengthened by ribs. The regulation of the water entering the combining cone is accomplished by moving

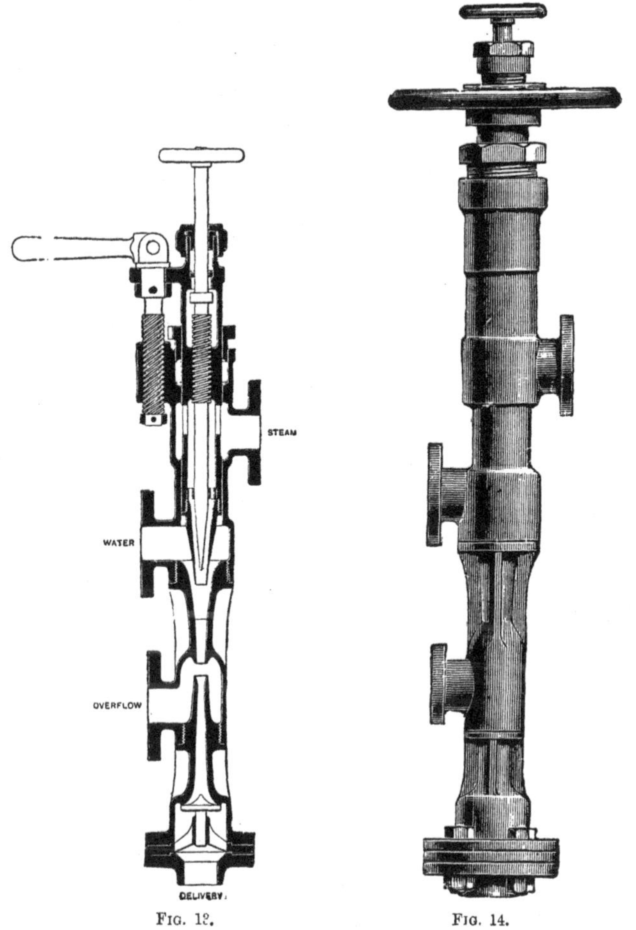


Fig. 13. Fig. 14.

the steam cone along its axis, and thus extending or

contracting the annular space between it and the combining cone. Two lugs are cast, one upon the hollow rod near its upper end, and another on the outer casing adjoining the stuffing box. In the former of these a spindle turns freely, having on its lower half cut five or six separate threads of the same pitch ; this screwed portion fitting into a similar screwed hole in the latter lug. By securing a washer on the screwed spindle just below the upper lug, the steam cone will receive the same axial motion as the screwed spindle. The object of the large number of threads on the water-regulating spindle is to complete the regulation instantly with only a partial turn of the handle.

We will assume it is required to start feeding a boiler by lifting the feed water from a well. In this form of apparatus there are no openings to the combining cone through which the steam can escape to the atmosphere, except through the usual overflow orifice, and therefore it will be absolutely necessary to use a steam orifice much smaller than the overflow during the time which elapses before the feed water reaches the combining cone. This is done by withdrawing the central spindle slightly by a partial turn of the hand wheel. The steam cone is then raised by turning the other handle. When the water appears at the overflow, the hand wheel is turned back so as to completely withdraw the plug spindle, thereby admitting a full supply of steam, or sufficient to drive the feed water into the boiler. Should either water or steam appear at the overflow after this, its particular regulator must be partly closed until it disappears. The illustration, fig. 13 (p. 357), has been kindly supplied by Messrs. Holden and Brooke. Most of the standard makers manufacture a similar injector. This pattern is called the "eccentric," to distinguish it from a slightly different variety called the "concentric," fig. 14, now nearly obsolete, though at one time made by Messrs. Sharpe, Stewart, and Co., who originally introduced the invention into England, soon after it was first brought out by M. Giffard. The above firm continued to carry on the manufacture of injectors without intermission for nearly a quarter of a century, until it was decided to remove their engineering works from Manchester to Glasgow, when they disposed of the whole of their injector business to the Patent Exhaust Steam Injector Company Limited, of Manchester, who now possess the original drawings and patterns, and to whom we are indebted for the illustration. These in themselves, from an historical point of view, must form an interesting collection. In passing, we may mention that it is a very strange fact that nearly all the leading injector makers hail from Manchester.

The only difference between figs. 13 and 14 is the method of raising the steam cone. In the "concentric" pattern a larger hand wheel is keyed to the hollow spindle which

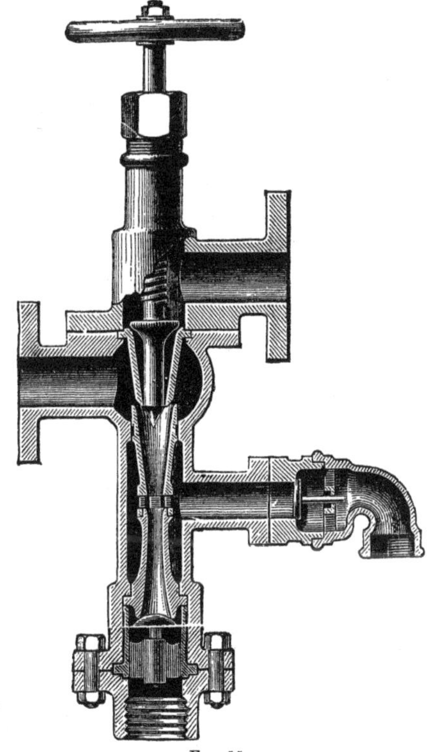

Fig. 15.

carries the steam cone. A thread is cut upon this hollow spindle, which screws into a companion thread cut in the casing.

Another form of water regulation which is sometimes used is obtained by constructing the combining and diverging cones similar to those in Mr. Webb's injector, fig, 10. and moving them axially by a rack and pinion situated in the overflow chamber.

A modified form of the original "Giffard," made by Messrs. Schäffer and Budenberg, is shown in fig. 15. The combining and diverging cones are screwed together, just below the overflow, and then inserted into the outside casing, being held there by the guide frame of the back-pressure valve. The outer casing is made in two parts, bolted together at the top of the steam cone. This fulfils two objects. The steam flange may be turned round at any time to suit the conditions of circumstances, without in any way impairing the action or usefulness of the apparatus ; and if the top casting is removed, the remainder forms a complete non-lifting injector. Of course, it is not necessary to metamorphose an injector in this way when once completed, but the same patterns and templates would do for the lifting as well as for the non-lifting types. The steam regulating spindle is tapered at its lower extremity for the purpose of graduating the admission of steam while the feed water is being lifted to the combining cone before starting ; and it also has another conical portion, which fits down upon the top of the steam cone, thus forming a stop valve for the steam supply. With this arrangement the steam valve on the boiler may be dispensed with. A non-return valve is also placed in the overflow pipe, preventing any air being taken into the boiler with the feed water. The overflow valve also augments the efficiency of the injector as a pump—a result which may be gathered from equation (44)—though the increase is small.

In many of the American types of injectors there is an overflow valve situated in the chamber beyond the diverging cone, which is opened or closed by the same lever that operates the steam valves. An example will be given when we come to deal with compound injectors.

Another method of exhausting the water pipe before the injector starts, which is much used in American injectors of the Sellers type, and on some others which are made in this country, is shown in fig. 15A. The steam regulating spindle has a small hole drilled for some distance along its axis at its conical end, terminating near its parallel part in two transverse holes. When it is desired to start the injector, the spindle is slightly withdrawn, so that steam may enter by the transverse holes, and issue into the combining cone by the longitudinal channel. The orifice of this channel being much smaller than that of the overflow, a vacuum is produced in the water pipe, as previously explained. On the arrival of the water at the overflow the spindle is withdrawn, so that the full amount of steam can enter the injector. The arrangement shown in the figure is taken from one of Messrs. Fairbairn and Hall's injectors. The same method has been employed by Messrs. Holden and Brooke, and others.

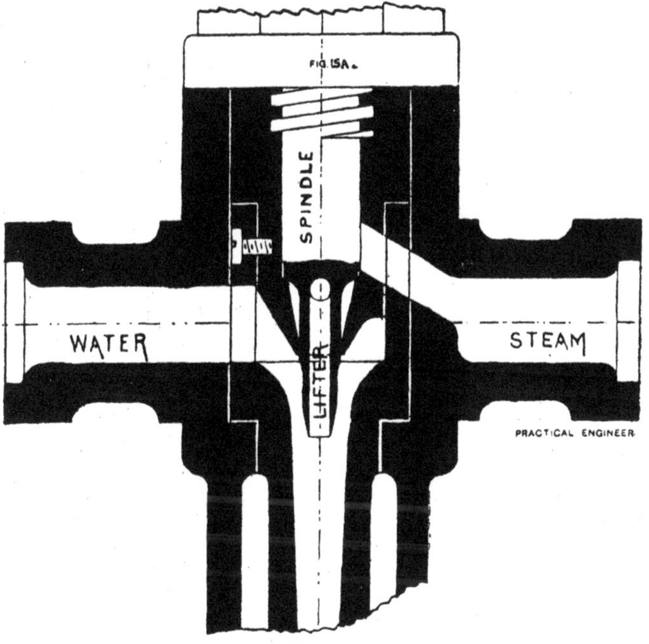


FIG. 15A.

It has been shown—equation (36)—that there is a given maximum temperature above which an injector will not take the feed water for a given pressure of steam. It has also been shown that variations in steam pressure affect the delivery and may cause it to cease working altogether. In the lifting injectors so far described this difficulty may be overcome to a certain extent by manipulating the steam and water supplies by the several means described; but injectors are now made in great numbers which are termed *automatic* or *re-starting*, from the fact that if they stop through inequalities in the road, or passing over points, they will immediately re-start again without any further regulation on the part of the attendant.

The first attempt at automatic regulation, so far as the author is aware, was that of Mr. Sellers, of the United

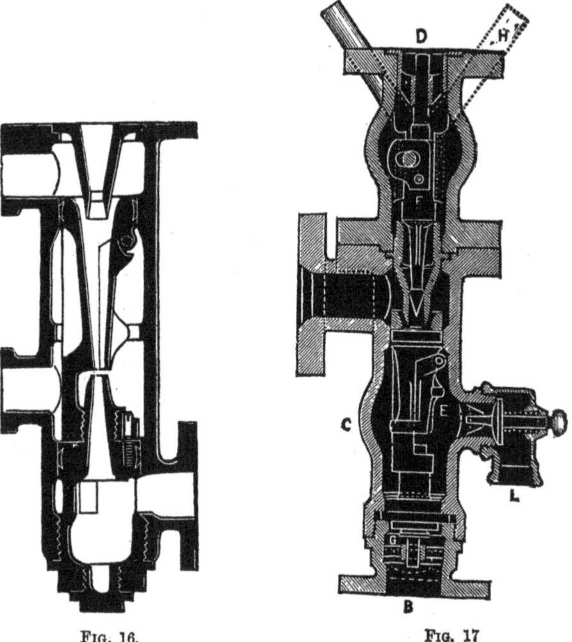


FIG. 16. FIG. 17

States, in 1866. The combining cone was fitted with an annular piston at its upper end, and was made to slide longitudinally by a variation in the steam pressure. A full detailed account of this injector, with illustrations, is to be found in the Minutes of Proceedings of the Institution of Mechanical Engineers, 1866. While this injector was designed to automatically regulate the supply of feed water, it in no way could start of itself after being accidentally stopped without a further manipulation of the steam spindle and valve, so that it could not lay claim to be classed as a re-starting injector.

In 1880 Messrs. Davies, Hamer, and Metcalf were engaged in perfecting an exhaust steam injector, in which the combining cone was split into two halves for a portion of its length by a plane passing through its axis; the piece which was thus split off being replaced and hinged to the main portion of the cone at the upper end of the split. Experiment showed this improvement to be so automatic in its action that it was attached to the ordinary patterns of non-lifting injectors, thereby converting them into *lifting automatic re-starting* injectors, which, after very slight modification, now appear as shown in fig. 16. This type is usually applied to locomotives. The reader will notice that the steam and diverging cones are of the same shape or profile as in the non-lifting pattern, but the method of securing the diverging and combining cones is somewhat different, the two being screwed together to the right of the overflow gap. The whole is then inserted into the outer casing, and secured to it by a few threads at the delivery end. The other joints are rendered tight by inserting packing in the double V grooves. The combining cone is cast in two pieces, the two being hinged together as shown in fig. 16. Perhaps a better idea may be gained of it from the elevation of it in fig. 17, at E. The two pieces are machined and fitted together, and then bored, so that the cone shall be perfect when the flap is closed. When steam is turned on, after issuing from the first cone, it enters the second and pushes up the flap, thereby securing a large area of outlet—much larger than that of the steam cone (an absolute necessity with full steam on)—which enables the water pipe to be exhausted, and the feed water raised to condense the outflowing steam. On the arrival of the cold feed water the steam is instantly condensed; at the same time the pressure in the overflow chamber closes the flap, and retains it closed until, by some cause or other, the water stream is broken, and the injector stops. The flap is then forced open, the feed drawn up, and the injector is at work again. The attendant working the injector always becomes cognisant of the injector starting to force water into the boiler by the distinctly audible clack as the flap closes up.

This instrument is specially well adapted for locomotives on account of its re-starting capabilities. A driver always has enough to do to look after the regulator, the reversing lever and signals, and the fireman finds that on the average express journey he has none too much time on his hands in which to be watching and manipulating the injector; so that a really automatic re-starting injector is often a great boon.

Another important feature in this injector is the facility with which the two water cones can be withdrawn for cleaning without breaking any pipe joints. In case the combining cone should stick and become unscrewed from the diverging cone, a little screw is inserted at the joint, so that they must both come out together. On account of the automaticity of action, virtually no water is wasted at the overflow at any time.

An interesting paper on this injector, together with the automatic exhaust injector, was read before the Institution of Mechanical Engineers in May, 1884, by Mr. A. Slater Savill, M.I.Mech.E., manager of the Exhaust Injector Co., who are the possessors of the patent rights.

---

## CHAPTER XI.

Messrs. Schäffer and Budenberg use the flap nozzle in their automatic re-starting injectors, under royalty to the Exhaust Injector Co. A section of their injector is given in fig. 17, and a perspective elevation in fig. 18. There are many noteworthy features to be found in this pattern, to which we will now devote our attention. As in the same firm's modified form of the original "Giffard," the lower part of the casting contains the whole of the cones—in fact, the injector proper; while the top casting is added for the sake of the regulating device, and can be turned round into four different positions, so that the same injector is at once a *right* and *left* hand pattern. The steam from the boiler passes through the top casting, which contains the regulating spindle F, which also carries the steam stop valve. This spindle, which increases the general efficiency of the apparatus, especially when working under abnormal conditions, is actuated by the hand lever H, which turns the little eccentric pin shown in the figure, the pin moving in a slot. The pin is so situated that it effectually locks the steam valve when closed. In the perspective view there will be noticed, attached to the handle, a graduated arc, which registers the correct position of the lever for different steam pressures by means of a fixed pointer; so that in starting the injector the handle has only to be moved over such that the index points to the then boiler pressure, and the injector easily starts. This it will do without any water regulation between the pressures of

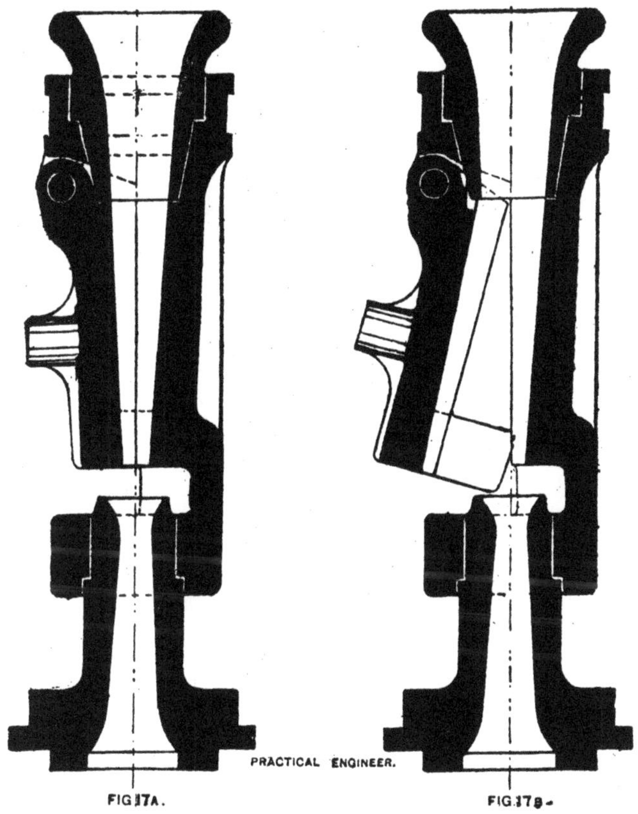


FIG. 17A. FIG. 17B.

30 lb. and 165 lb.; below this limit a little water regulation may be necessary. The regulation by means of a handle in preference to the screw was adopted on account of the simplicity of manipulation with the former.

A non-return valve is generally used on the overflow orifice, and kept on its seating by a weak spring, the valve stem passing through the casing, so that it may be made to fit tight if by sediment or other means it should become tiled. On the locomotive pattern there are more overflow orifices than one, the remainder being covered with blank flanges. This allows of the body of the apparatus being

Fig. 18.

turned into any position when necessary. The casing is bellied out at the overflow so as to act as an air space round the flap, which assists in securing good working. So that the water in the tender of a locomotive may be warmed by surplus steam, a screw-down overflow valve is placed on one of the pair of injectors. The non-return overflow valve, besides increasing the efficiency under ordinary circumstances, causes it to work dry under certain extreme conditions, such as high temperature feed water. It also prevents loss of water at re-starting, as it closes at the instant condensation begins; and the momentary discharge of water collects in the overflow chamber, and is drawn into the boiler.

A detail of the diverging and improved combining cone is shown in fig. 17A and fig. 17B. In the former the flap is shown closed, exactly as it is when the injector is at work; while the latter shows the flap open, as it would be when

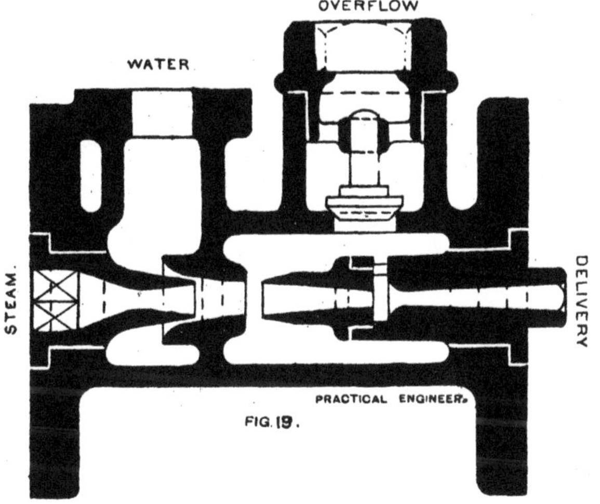

FIG. 19.

the injector is idle and placed in a vertical position. On the end of the flap, remote from the hinge, are cast a pair of ears, which act as guides should the hinge become at all worn. The upper portion of the combining cone is let into the main portion, so as to secure the continuity of the jet. The author has had the opportunity of examining some of these flap nozzles minutely, and he feels bound to express his admiration for the highest essence of workmanship exhibited throughout their construction. The flap fits upon the main part so well, that the joint is hardly perceptible when closed, and quite air-tight.

The following will indicate what injectors of this type will do with the water valve full open. With steam at 60 lb. per square inch it will lift 18 ft. with feed water not exceeding 105 deg. Fah., or lift the same distance at 180 lb. per square inch with feed water at a temperature not exceeding 80 deg. Fah. With a pressure ranging between 30 lb. and 60 lb., and feed water at 140 deg, Fah., the injector will still lift a height of 3 ft. without regulation of feed-water. With no lift, and steam at 180 lb., it will take feed at the high temperature of 105 deg. Fah. This is a temperature over 20 deg. higher than that given as a maximum in Table IV., which confirms the author's suggestion, made in another part of these pages, that, when

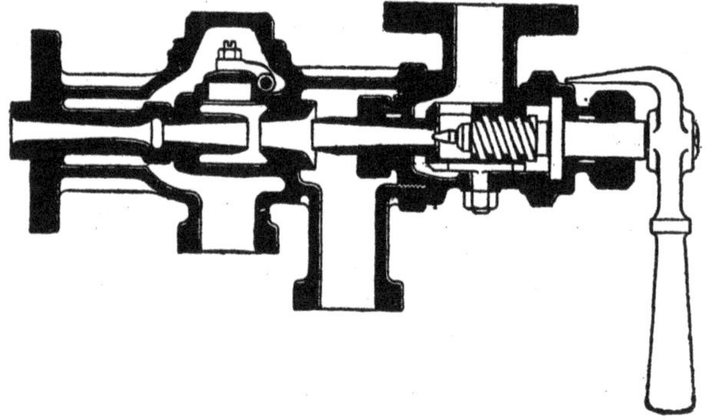

Fig. 20.

dealing with feed waters at high temperatures, the whole of the steam is not condensed in the combining cone, although it may be so after it has passed the overflow and diverging cone orifice, and becomes subject to higher pressures. The overflow valve also assists the injector to feed at higher temperatures. The injector starts or re-starts perfectly and absolutely without loss of water. It can be placed either vertically or horizontally, but when in the latter position it must never have the flap underneath.

The author does not attach so much importance to the capacity of an injector to lift a great height as to its automaticity in re-starting, for it is seldom an injector is required to lift more than 5 ft. or 6 ft., while the necessity for its automatic qualities is evident.

At the discussion on Mr. Savill's paper, previously alluded to, some engineers suggested that sediment deposited from water containing lime, magnesia, or aluminium in solution would materially affect the efficiency of the injector; but examples of nozzles were shown at the same meeting which had been working with water anything but pure, and had not been removed from the casing for over two years, working well the whole of the time.

Mr. John H. Hosgood, M.I.Mech.E., chief mechanical engineer of the Barry Docks and Railway, informs the author that the flap-nozzle injectors which he has had fitted to one side of most of the locomotives under him (the other side being fitted with the same firm's—the Exhaust Injector Co.—exhaust injector) do not require examining any more

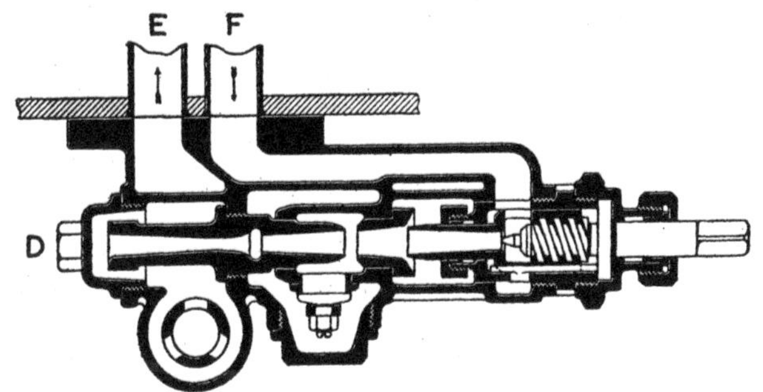


FIG. 21.

often than any other pattern of injector, and they work excellently, although the water used contains an enormous amount of lime and magnesia in solution; so much so that his company have gone to the expense of bringing the water used for steam purposes a distance of some miles.

Another form of re-starting injector is shown in fig. 19. The patent for this was taken out by Mr. C. S. Madan so late as 1891, and it is the standard type now made by that firm. The object sought after in this particular design is common to all re-starting injectors, namely, the provision of an exit area for the steam somewhere in the combining cone, so that a partial vacuum may be produced in the water pipe to allow the atmosphere to force up the feed water to the combining cone to start the apparatus. This is here accomplished by taking an ordinary simple non-lifting instrument, and cutting a piece out of the combining cone near its middle by a couple of planes perpendicular to the axis. So large an aperture in the region of mixture of steam and feed water would permit of an excessive quantity of air being carried into the boiler, which is not only detrimental to the working of the engine and boiler, but breaks up the column of water passing through the injector, an event generally accompanied by its ceasing to work. This difficulty is easily surmounted by placing a non-return valve on the overflow orifice in the outer casing. The valve may or may not have a weak spiral spring to keep it closed.

It seems strange that a large gap can be made in the combining cone right in the neighbourhood where we should imagine that the disturbance caused by the inrushing steam jet to be great, and where the water and steam would be most likely to escape. Such, however, is not the case, as the injector works without loss of water. Should a little water escape from that opening, it will be carried into the boiler through the usual overflow orifice.

It seems as if the major portion of the steam is instantly condensed on coming in contact with the feed water, and that the pressure of the steam, after issuing from the steam cone, immediately falls to that of the overflow chamber, or very near it. The pressure of vapour in that chamber, together with the tendency to contraction of the jet, enables it to leap over the first overflow space. The first part of the combining cone is cast in the outer casing; the remainder are screwed into their respective positions.

We next come to an injector which is unique, in so far that it is perfectly automatic in re-starting, and at the same time contains in itself both water and steam regulator, operated by a single handle. A longitudinal section of this injector, which is the invention of Mr. R. G. Brooke, of the firm of Messrs. Holden and Brooke Limited, is shown in fig. 20. Its capacity for restarting automatically is derived from the transversely divided combining cone (as in the last example), the outlet to the little chamber surrounding this division being closed by a flap valve, while the ordinary overflow orifice is in communication with the atmosphere, as usual.

The action of the flap valve is precisely similar to that of the flap nozzle. Of course, an ordinary disc valve may be used, as in the previous case. The makers have made the seating of the flap so that it may be turned round into any position at a moment's notice if necessary, thus allowing the valve to be so placed that it will always fall upon its seat, this being its most efficient position. The exterior surface of the steam cone is cylindrical, and slides axially, without turning, in the stuffing box and gland, as shown in the figure. The upper portion of this cone has apertures through which steam may pass to the central channel, the opening to which is regulated by the coned spindle, which also carries the stop valve. In the prolongation of the upper part of the cone are cut some threads of a quick pitch, into which gear corresponding threads cut on the spindle, the cone being prevented from turning round by the feather key fixed in the outer casing. A collar on the

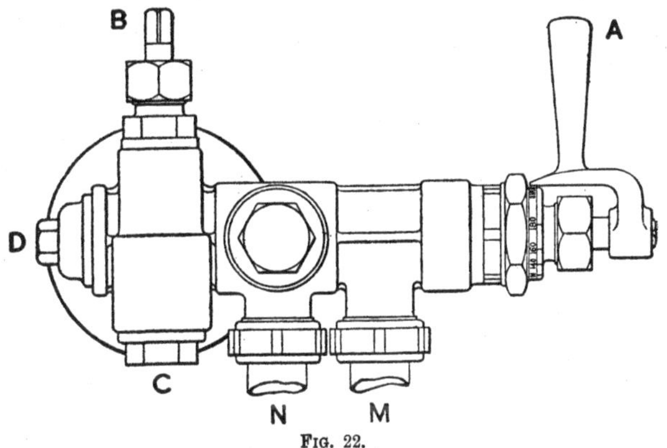


Fig. 22.

spindle prevents its axial motion. The threads being of a large pitch, only a partial turn of the handle is required to render the passages full open. It has been shown—equations (52) and (53)—that when the diverging cone is of constant sectional area, that for varying steam pressures, the steam orifice should decrease as the feed-water orifice increases, and *vice versa*. (A further discussion of this point will be found in the appendix.) This is admirably brought about by a partial turn of the steam spindle. Also it has been shown that an increase in pressure must be accompanied by a decrease in the steam cone orifice for the most efficient working. The lever or handle is prolonged beyond its boss, tapered, and bent over, the extremity pointing to a series of numbers on the cylindrical part of the nut, which indicate the pressure of steam for that particular position of the handle pointer. By this method of regulation the injector works equally well with almost any variation of steam pressure. It will thus be seen that by a single movement of the handle the steam is turned on, the injector started, regulated for any steam pressure, and the quantity and temperature of feed water (as delivered to the boiler) varied between the two extremes of maximum and minimum. The moving parts are situated in the steam space, so they are not liable to become deranged through deposit from dirty water. The flap valve does not

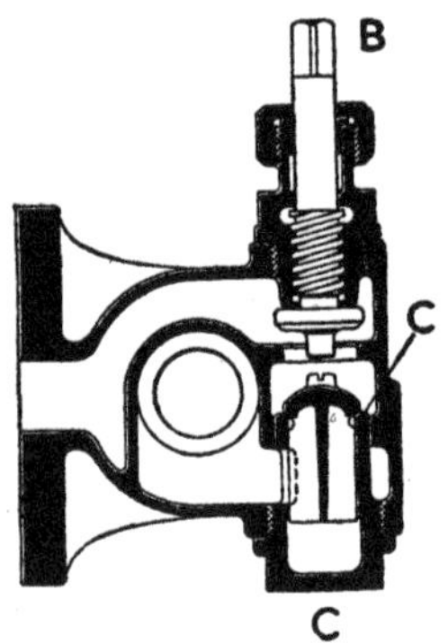

FIG. 23.

come in contact with any water, and hence the seating is not affected with sediment.

Of recent years it has become a custom with many locomotive engineers to place the feed clack box at the back of the boiler, so that it can be examined from the foot plate, the feed being carried into the front end of the boiler by an internal pipe. This arrangement has still further been improved by combining with the clack box the steam valve, steam and delivery pipe flange, injector, and screw-down valve. The makers of the injector just described (the "1890" pattern) have adapted it to the combination form shown in figs. 21, 22, 23, and 24, and known as the "1892" combination pattern. It is fitted either on the back or side of the firebox, inside the cab. There is an internal steam pipe to bring dry steam from the dome or manhole to the injector, and the stop valve on the regulating spindle is the only one necessary. Not only can the cones be removed from the casing while the engine is under steam, but the clack valve can also be removed at the same time, by first securing the screw-down valve, fig. 23, and then removing the nut underneath it. In this manner the valve, or its seating, can be taken out and turned or re-ground without affecting the remainder of the casting.

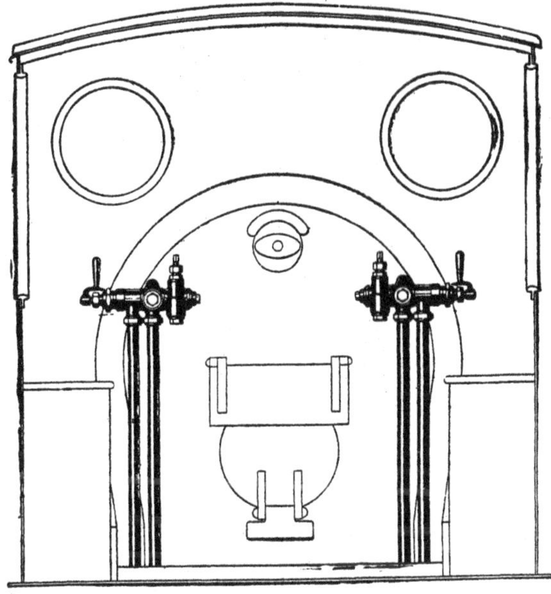

FIG. 24.

Another combination injector, and one which is used on English railways to a greater extent at the present time than any other *combination* injector, is that of Messrs. Gresham and Craven, two longitudinal sections of which are given in figs. 25 and 26. In this form it has remained, with slight modification, since it was brought out in 1884. It is arranged vertically instead of horizontally, as in the last instance, which makes it slightly neater in appearance. Steam is admitted through the lower channel in the flange

R
D
DELIVERY
D
Y
STEAM
A
H
B
H
C
O
H
S
P
W
W
PRACTICAL ENGINEER
OVERFLOW
WATER

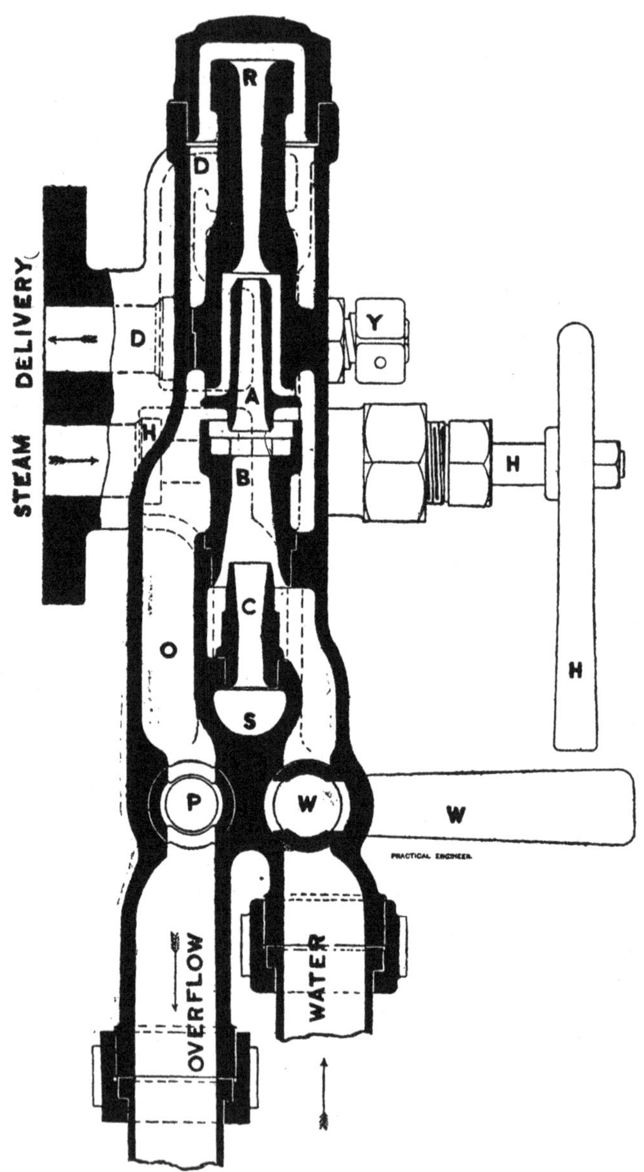

FIG. 25.

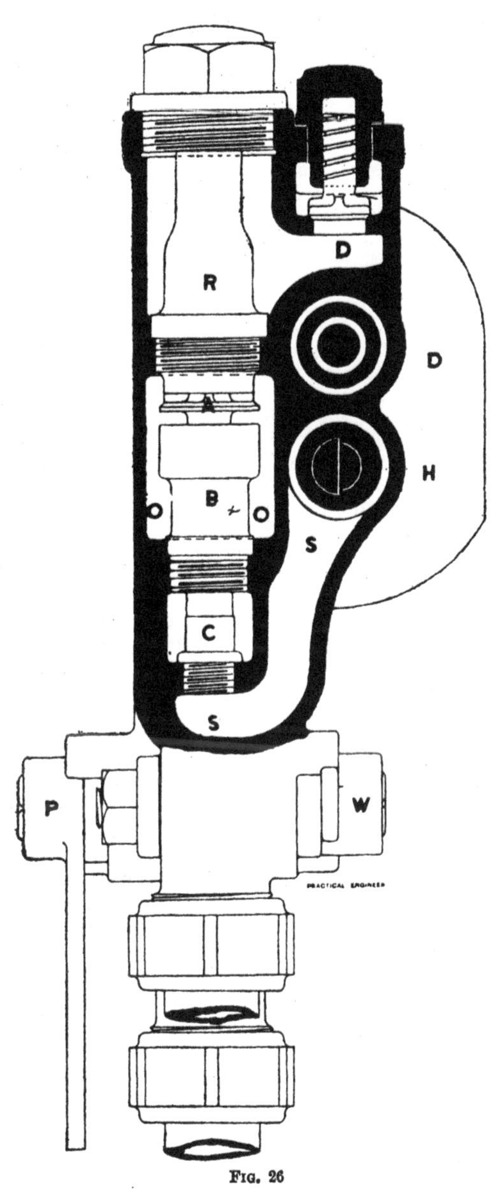


Fig. 26

by the steam valve H, from which it passes to the steam cone C, through the channel S. The non-return screw-down valve Y covers the delivery orifice, in case the clack valve (in the left-hand section) should become tiled or the cones require to be removed under steam. The latter can be accomplished by first taking off the nut just above the cone R, and then unscrewing each cone in turn from the main casting. There are hand valves, W and P, on the feed-water and overflow pipes respectively; the former for regulating the supply of water; the latter for heating the feed water with surplus steam. The principal feature of this apparatus is the device from which it derives its automatic re-starting property.

In the lower portion of the diverging cone is the small loose cone A, capable of sliding axially, its axis being maintained coincident with that of the other cones by four wing guides, three of which are shown in the fig. 26. In the figure, the loose cone is shown in the position it occupies when the injector is exhausting the water pipe, the steam being able to escape into the overflow channel O, through the space between the combining cone B and the loose cone A; and also at the end of the latter, the wings preventing the flange of the cone A from touching the diverging cone. Directly the water arrives and condenses the steam, the partial vacuum formed in the combining cone allows the loose cone to fall until its flange rests upon the combining cone, it being assisted and held there by the pressure of the atmosphere from without. In this closed position the two cones A and B together form one long continuous cone.

This injector works quite dry, and is very much liked by drivers and firemen. It is with this object in view that it has been placed on many engines running trains through suburban districts of large towns, the continual blowing through of steam when stopped at long platforms being objectionable with trains that stop so often.

A peculiar feature with all combination injectors is that there are no pipes under pressure, except when heating the feed water.

# CHAPTER XII.

## Exhaust and Compound Injectors.

The fact that the ordinary live steam injector would force feed water into a boiler whose pressure of steam was considerably in excess of that from which the injector derived its motive power, was received by engineers with no little incredulity when the invention was first brought before the public, seems anything but unreasonable; especially when it is borne in mind that at the period of its introduction, about thirty years ago, a knowledge of hydro-dynamics amongst engineers was a rarity; while the science of thermo-dynamics was in its infancy, though being rapidly developed by such profound thinkers as Rankine, Kelvin, Joule, Clausius, and Regnault.

Various theories were advanced to account for the injector's action, nearly all of which had not the slightest reference to philosophic reasoning. While this state of things prevailed, can anyone wonder that the idea of using steam for the motive power, after it had previously done work in the cylinder of an engine, was not carried into effect—perhaps not even thought of seriously?

A reference to the ninth column of Table III. shows that the velocity of steam issuing from an orifice under widely different pressures does not vary very considerably for maximum weight flow; for instance, the velocity under a pressure of 15 lb. per square inch is, roughly, 1,340 ft. per second for maximum weight flow; while the velocity at 100 lb. per square inch is about 1,420 ft. per second. The difference between these two quantities is small compared with either of them, so that for rough purposes they may be considered the same. Now, the power of an injector to force water against a pressure is derived from the momentum of the working steam destroyed per second. This momentum is the product of the mass of the steam and its velocity, and the mass is that quantity which passes a given point in a unit of time. This clearly depends upon the velocity of the steam, the area of orifice, and the density. Hence the momentum destroyed per second equals the product of the density, the area of orifice, and the square of the velocity. Above, the velocities of efflux of steam at the pressures of 100 lb. and 15 lb. per square inch were roughly the same; hence the power of two injectors, using steam at 100 lb. and 15 lb. respectively, would be directly as the product of the steam, orifice area, and density of steam at 100 lb. is to the orifice area and density of steam at 15 lb. If the orifices are made inversely as the densities, we shall have two injectors of equal power. There are practical difficulties to be overcome in carrying out this idea in accordance with mechanical principles, and there are also thermo-dynamical difficulties which prevent us from realising in practice this ideal state of things, so that we are unable to construct an exhaust injector that will deliver water against such high pressures as a live steam injector; but although we cannot do that, we can use an exhaust injector to force water against intermediate pressures, which in some cases approach within a reasonable distance of those which are overcome by a live steam apparatus.

To what extent we may hope to accomplish this end we now propose to show. In what follows, the influence of the pressure of the atmosphere, or the head of feed water, will be neglected.

In our investigation of the live steam injector, we saw that as the temperature in the combining cone never exceeded 212 deg., the pressure there could not be greater than about 15 lb. per square inch; a quantity always less than three-fifths of the boiler pressure. Also when this pressure was equal to, or less than, three-fifths of the pressure of the boiler steam, the maximum weight flow of steam occurred.

Now, with the exhaust injector, the pressure corresponding to the boiler pressure in the live steam instrument is 15 lb. per square inch, while the pressure in the combining cone may be anything between zero and 15 lb. per square inch, according to the temperature of the mixed water and steam. If this latter pressure is equal to, or lower than, 9 lb., we shall have the maximum weight flow of steam occur, and its velocity of efflux will be given by equation (10) or equations (22) or (23), preferably the latter. Should the amount of feed water be comparatively small, such that the temperature of the combined steam and water would cause a pressure greater than 10 lb. per square inch, we must return to equation (14) for the velocity of efflux of the exhaust steam into the injector; in which we shall have $P_1 = 144 \times 15$ lb. $= P_a$ say, and $P_2 =$ the pressure per square foot in the combining cone $= P_3$. We may also notice here, that when $P_3$ is great, comparatively speaking, the velocity of the issuing steam is correspondingly small; and, necessarily, the velocity $v_3$ of the mixed steam and water must be small. Again, if the pressure in the combining cone is small, the temperature there must also be correspondingly low; which would be produced by a large quantity of feed water per pound of steam. This would also produce a small velocity $v_3$ of the combined water and steam. Our object, then, is to find the proportion of water to steam which will give the maximum velocity $v_3$ to the combined steam and water.

As before, let $\beta$ be the number of pounds of feed water per pound of steam, and $v_a$ the velocity of efflux of the exhaust steam. Then, from the equality of momenta, we have, as in equation (30),

$$v_3 = \frac{v_a}{1+\beta} \quad . \quad . \quad . \quad . \quad . \quad . \quad . \quad . \quad (54^a).$$

Now, from equation (14) we have—

$$v_a = v_2 = \sqrt{2\,g\,\frac{n}{n-1}\left(\frac{P_a}{\rho_a} - \frac{P_3}{\rho_3}\right)} \quad . \quad . \quad . \quad (55)$$

Also we have $\quad P = K\,\rho^n$

therefore $\quad \left(\frac{P}{K}\right)^{\frac{1}{n}} = \rho$

and $\quad \frac{P}{\rho} = \frac{P}{\left(\frac{P}{K}\right)^{\frac{1}{n}}} = K^{\frac{1}{n}}\,P^{\frac{n-1}{n}} \quad . \quad . \quad . \quad . \quad . \quad . \quad (56)$

Substituting in (55) with suffixes, we find

$$v_a = \sqrt{2\,g\,\frac{n}{n-1}\,K^{\frac{1}{n}}\left(P_a^{\frac{n-1}{n}} - P_3^{\frac{n-1}{n}}\right)} \quad . \quad (57)$$

The value of $v_a$ thus determined depends to a very great extent upon the value given to $n$. That value depends upon the quality of the steam; and in the exhaust pipe of an engine it probably varies considerably under different circumstances. Thus, in fixing a value for $v_a$ we must allow a margin for variation of circumstances. Should the exhaust steam be wet, as it often is, then the dry steam has to do more work than it would have if the whole of the steam were dry; and therefore the velocity of the exhaust steam will be reduced in proportion to its wetness. We may obtain an approximation to the velocity $v_a$ by assuming the steam dry and $n = \frac{10}{9}$. in equation (57), where the most convenient form for calculation is—

$$\log v_a = 3{\cdot}6106579 + \frac{1}{2}\log\left(2{\cdot}15499 - 144^{\frac{1}{10}}\,p_3^{\frac{1}{10}}\right) \quad . \quad (58)$$

In this we have assumed the pressure of the atmosphere 15 lb. per square inch.

The author has calculated several different values of $v_a$, corresponding to different values of $P_3$, from which the curve number 1, fig. 27, has been plotted.

Now to obtain the amount of feed water per pound of steam that will give the combined jet its maximum velocity. For steady motion equation (39) must hold. Neglecting $v_4$, the velocity with which the jet enters the boiler, we have

$$\frac{P_4 - P_3}{\rho} = \frac{v_3^2}{2g}$$

and by replacing the right-hand side by its equivalent in equation (54$a$) we have

$$2\,g\,\frac{(P_4 - P_3)}{\rho} \quad \left(\frac{v_a}{1 + \beta}\right)^2$$

For $v_a$ we may substitute its value in (57), and the above then becomes

$$2\,g\,\frac{P_4 - P_3}{\rho} = \frac{n}{n-1}\quad\frac{K^{\frac{1}{n}}}{(1+\beta)^2}\left[P_a^{\frac{n-1}{n}} - P_3^{\frac{n-1}{n}}\right] \quad . \quad (59)$$

in which we have three variables, viz., $(P_4 - P_3)$, $(1 + \beta)$, and $P_3$. It will now be more convenient to replace the two last by their equivalents in terms of the temperature $t_3$.

Rankine has given us the following accurate relation between temperature and pressure:

$$\log_{10} P = A - \frac{B}{T} - \frac{C}{T^2} \quad . \quad . \quad . \quad . \quad . \quad (60)$$

in which P = pressure in pounds per square foot, A = 8·2591, T = the absolute temperature Fah. of the vapour (saturated steam), B = 2731·6, and C = 396,945; $\log_{10} B = 3{\cdot}43642$, and $\log_{10} C = 5{\cdot}59873$. (*Vide* "The Steam Engine and other Prime Movers," by Rankine, p. 237.)

Multiply both sides of equation (60) by $\mu = 2{\cdot}30258$, the modulus for converting common to Napierian logarithms, and we have

$$\log_e P = \mu \left[ A - \frac{B}{T} - \frac{C}{T^2} \right]$$

and, after reversing the operation of taking logarithms, we find that

$$P = e^{\mu \left[ A - \frac{B}{T} - \frac{C}{T^2} \right]}$$

nd, consequently,

$$P^{\frac{n-1}{n}} = e^{\mu \left(\frac{n-1}{n}\right) \left( A - \frac{B}{T} - \frac{C}{T^2} \right)}$$

After substituting in (59),

$$2\,g\,\frac{P_4 - P_3}{\rho} = \frac{n\,K^{\frac{1}{n}}}{n-1} \left( \frac{T_3 - T_5}{1178 - t_5} \right)^2$$

$$\left[ e^{\mu \left(\frac{n-1}{n}\right) \left( A - \frac{B}{Ta} - \frac{C}{Ta^2} \right)} - e^{\mu \left(\frac{n-1}{n}\right) \left( A - \frac{B}{T_3} - \frac{C}{T_3^2} \right)} \right] \quad . \quad (61)$$

To find what value of $T_3$ will make this expression a maximum, differentiate the right-hand side with respect to $T_3$, and equate the first differential coefficient to zero. The temperature $T_5$ is constant. We then find that—

$$\mu \left(\frac{n-1}{2\,n}\right) \left(T_3 - T_5\right) \left( \frac{B}{T_3^2} + \frac{2\,C}{T_3^3} \right) e^{\mu \left(\frac{n-1}{n}\right) \left( A - \frac{B}{T_3} - \frac{C}{T_3^2} \right)}$$

$$= \left[ e^{\mu \left(\frac{n-1}{n}\right) \left( A - \frac{B}{Ta} - \frac{C}{Ta^2} \right)} - e^{\mu \left(\frac{n-1}{n}\right) \left( A - \frac{B}{T_3} - \frac{C}{T_3^2} \right)} \right]$$

Divide both sides of this equation by the latter term of the right-hand side, and we have—

$$\mu \left(\frac{n-1}{2n}\right) (T_3 - T_5) \left( \frac{B}{T_3^2} + \frac{2\,C}{T_3^3} \right)$$

$$= \frac{e^{\mu \left(\frac{n-1}{n}\right) \left( A - \frac{B}{Ta} - \frac{C}{Ta^2} \right)}}{e^{\mu \left(\frac{n-1}{n}\right) \left( A - \frac{B}{T_3} - \frac{C}{T_3^2} \right)}} - 1$$

$$= e^{\mu \left(\frac{n-1}{n}\right) \left[ B \left( \frac{1}{T_3} - \frac{1}{Ta} \right) + C \left( \frac{1}{T_3^2} - \frac{1}{Ta^2} \right) \right]} - 1.$$

But

$$e^x = 1 + x + \frac{x^2}{1 \times 2} + \frac{x^3}{1 \times 2 \times 3} + \text{etc., } \textit{ad infinitum},$$

and if $x$ is a small fraction, all terms beyond the second can be neglected with respect to those before it. The exponent of $e$ in the above equation cannot be greater than ·0909, and therefore in the expansion we can safely neglect all terms after the second, without introducing an error of more than one-third of one per cent. The above then becomes—

$$\frac{(T_3 - T_5)}{2}\left(\frac{B}{T_3^2} + \frac{2C}{T_3^3}\right) = B\left(\frac{1}{T_3} - \frac{1}{T_a}\right) + C\left(\frac{1}{T_3^2} - \frac{1}{T_a^2}\right) \quad (62)$$

If we assume that $T_5 = 520$ deg.—*i.e.*, $t_5 = 60$ deg. Fah.—and put in the values of B and C given above, together with $t^a = 212$ deg. Fah., the last equation reduces to—

$$\cdot 00363\, T_3^3 - T_3^2 - 1110\, T_3 + 150000 = 0 \quad . \quad . \quad . \quad (63)$$

a solution of which is found when $T_3$ is put equal to 648 deg. Fah. *absolute*, which is equivalent to about 187 deg. Fah.

It was mentioned just above that the right-hand side of equation (62) may be slightly in error by the approximation there used. If we increase the coefficient ·00363 in equation (63) by ·003 of itself, the amount of probable error, it becomes ·00364, which value would make $t_3 = 185$ deg. Fah. approximately. This we may then take as the temperature of the water as it enters the boiler (when supplied to the injector at 60 deg.) when moving with its maximum velocity. The corresponding amount of feed water per pound of steam is given by the equation—

$$1 + \beta = \frac{1178 - t_5}{t_3 - t_5}$$

from which we find that $\beta = 7{\cdot}9$ approximately.

Now, the velocity of the steam in the steam cone given by the curve No. 1, fig. 27, when the temperature in the delivery cone is 185 deg., is 1370 ft. per second. Putting this in equation (54*a*), we have—

$$v_3 = \frac{v_a}{1 + \beta} = \frac{1370}{8{\cdot}9} = 154 \text{ ft. per second.}$$

If there were no resistances in the injector or its

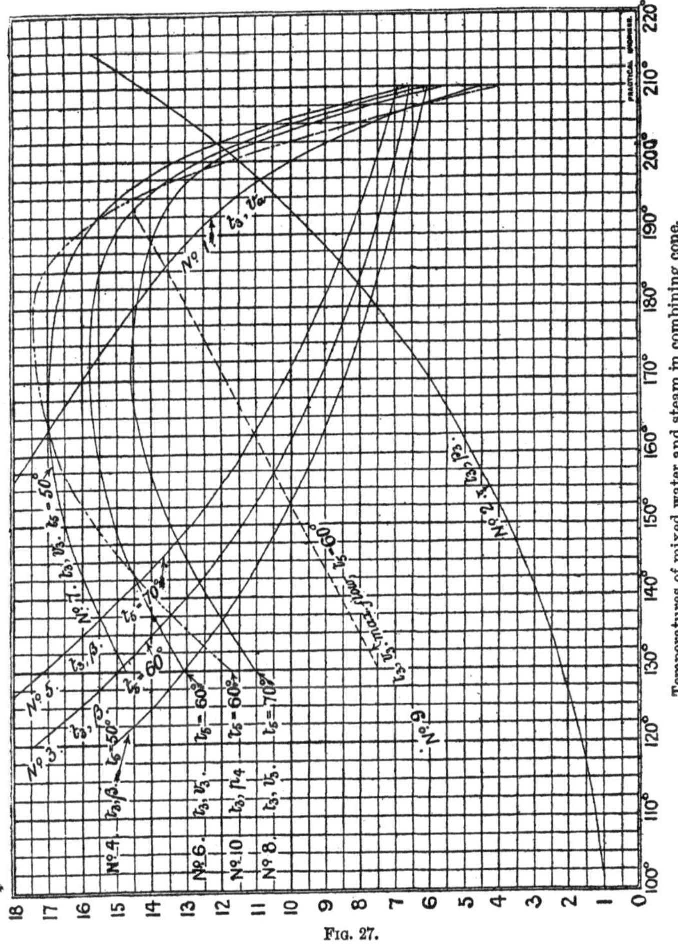


Fig. 27.

Maximum pressures in tens of pounds per square inch, against which the injector will feed.
Velocities in hundreds of feet per second of the exhaust steam.
Pounds of feed water per pound of steam.
Pressures per square inch in combining cone.

connections (especially the latter), this would force water against the pressure $P_4$ given by equation—

$$\frac{P_4 - P_3}{\rho} = \frac{v_3^2}{2g}.$$

From the curve No. 2, fig. 27, showing the relation between temperature and pressure, we may find $P_3$. We then find that $p_4 = 171$ lb. per square inch. The delivery water often takes a circuitous route from the injector to the boiler, passing through chambers and orifices, and by valves, whose construction and form are of the very opposite to that which the principles of hydraulics would indicate; and hence we must expect a loss of head between the delivery cone and the boiler. This loss of head sometimes amounts to as much as 60 per cent of the total head. The pressure against which the exhaust injector will generally feed in practice is about 70 lb. per square inch; though with a little manipulation it is possible to reach a higher pressure, as the following results of actual tests will show:—

Table VI.

| $t_5$ | $p_4$ | $t_3$ | $t_3 - t_5$ | $\beta$ | No. of Injector. | |
|---|---|---|---|---|---|---|
| 54 | 84 | 156 | 102 | 10 | 4 | |
| 81 | 55 | 182 | 101 | 9·8 | 4 | |
| 50 | 65 | 180 | 130 | 7·7 | 4 | This injector would not start when $t_5$ was greater than 75. |
| 60 | 105 | — | — | — | 10 | A little water issued from the overflow. |

The above results are collected from a paper read before the Owens College Engineering Society, by Mr. George Harrison, the injectors used being made by the Patent Exhaust Injector Company, Manchester.

In the fourth experiment the whole of the quantities were not taken, but supposing we assume that the quantity of feed water was such as to give a maximum velocity to the delivery jet, we should find that $\beta = 7·9$ approximately, and the temperature of the jet within 20 deg. of the ordinary boiling point.

The chief causes which tend to make the injector feed against lower pressures than those indicated by calculation are radiation, fluid friction, eddies, wetness of exhaust steam, and the resistance of chambers whose forms change abruptly, all of which have been neglected in the calculation. It may be roughly estimated that an injector will feed against a pressure of about 60 per cent of that indicated by calculation when the resistances are neglected therefrom.

A series of curves have been plotted in fig. 27, showing the relation between the several quantities relating to the working of the exhaust injector when the velocity of the delivery jet is a maximum. The abscissæ are for all the curves, the temperature in degrees Fahrenheit of the combining cone beginning at the left-hand side with 100. Curve No. 1 shows the velocity of efflux of the exhaust steam (supposed to be dry) when the pressure in the combining cone is that which corresponds to the temperature denoted by the abscissæ. For instance, if the pressure in the combining cone is 9 lb. per square inch, we find from curve No. 2 that the temperature corresponding to this pressure is about 188 deg., and then from curve No. 1 that the velocity of efflux of the steam about 1,310 ft. per second. The digits of the hundreds only are given on the left of the diagram, so that the same figures will serve for pounds per square inch in the pressure temperature curve, and for hundreds of feet per second in the velocity temperature curve.

Again, by putting $t_5 = 60$ deg. in the equation

$$1 + \beta = \frac{1178 - t_5}{t_3 - t_5}$$

and then giving $t_3$ successive values corresponding to the abscissæ, we get a series of values for $\beta$ corresponding to these temperatures of delivery. Then, by using the numbers to the left of the figures as pounds of feed water per pound of steam, we have curve No. 3 connecting $\beta$ with $t_3$. Two more similar curves, Nos. 4 and 5, are obtained by putting $t_5$ equal to 50 deg. and 70 deg. respectively. Now, by taking the several values of $\beta$, given by the three last curves, and the value of $v_a$ in curve No. 1, and substituting them in the equation

$$v_3 = \frac{v_a}{1 + \beta},$$

we obtain three other curves, Nos. 6, 7, and 8, giving the velocities of the jet in the delivery cone for different values of $\beta$ and given values of $t_5$. If we assume that the velocity of efflux of steam is that of maximum weight flow, then $v_3$ is given by the curve No. 9.

Again, by taking $t_5 = 60$ deg., and using the velocity given by curve No. 6, we may obtain the maximum ideal pressure, $p_4$ lbs. per square inch, against which it is possible for the injector to feed after substituting in the equation

$$\frac{P_4 - P_3}{\rho} = \frac{v_3^2}{2g}.$$

In this manner curve No. 10 has been plotted, and it will be noticed that it indicates a maximum pressure $p_4$ at a temperature of the combining cone, of about 180 deg.

---

## CHAPTER XIII.

### Compound Injectors.

We have seen in the previous pages that the ordinary injector derives but little assistance from the head of feed water, because of its insignificance; and that the temperature of the delivery jet cannot exceed about 200 deg. Fah.

Now, if we work a pair of injectors in series—*i.e.*, the delivery pipe of one being connected to the water pipe of the other—we shall then have feed water supplied to the second injector under a considerable head, and at the same time the temperature of the combined water and steam in

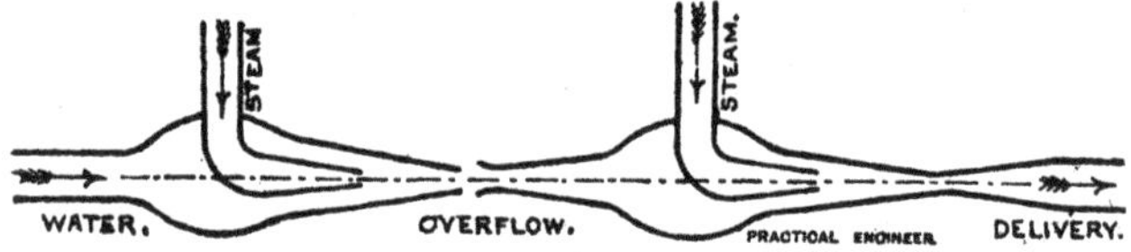


Fig. 28.—Ideal form of Compound Injector.

the second combining cone may be much above 200 deg., on account of the extra pressure. The ideal form of this injector is shown in fig. 28. That part of the apparatus to the left is called the "lifter," that to the right the "forcer." The more compact and actual arrangement of this form of instrument is shown, fig. 29, in which the two portions have their axes parallel to one another, and a steam pipe having two flanges supplying steam to both portions.

It is obvious that an injector of this construction is capable of forcing water into a boiler which carries a higher pressure than an ordinary single injector.

Let $\beta_2$ be the ratio of water to steam in the forcer, while $\beta_1$ represents the same ratio in the lifter; then for every pound of steam that issues from the "forcer" steam nozzle there are $(1 + \beta_2)$ lb. of water and steam that enter the boiler. Of the $\beta_2$ lb. of water in the forcer, we shall have $\left(\frac{1}{1+\beta_1}\right)$th part steam (condensed) and the remainder cold feed water; *i.e.*, $\frac{\beta_2}{1+\beta_1}$ lb. condensed steam and $\frac{\beta_1\beta_2}{1+\beta_1}$ lb.

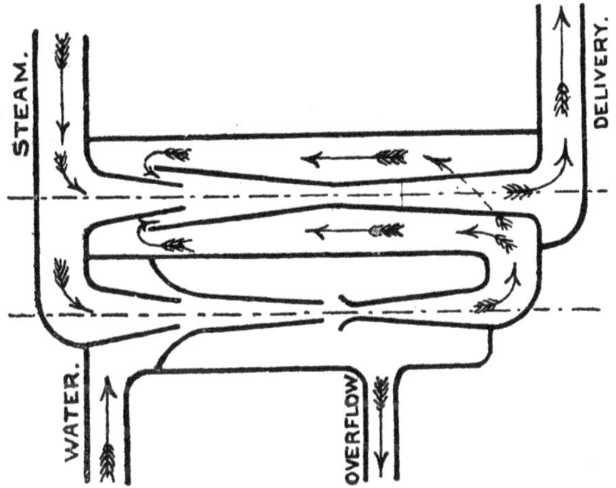


FIG. .—Actual form of Compound Injector.

of feed water. Therefore the total amount of steam used in both forcer and lifter per pound of steam in the forcer

$$= 1 + \frac{\beta_2}{1+\beta_1} = \frac{1 + \beta_1 + \beta_2}{1 + \beta_1} \text{ lb.} \quad . \quad . \quad . \quad . \quad . \quad . \quad (64)$$

Net amount of feed water per pound of steam in the forcer

$$= \frac{\beta_1\beta_2}{1+\beta_1} \text{ lb.} \quad . \quad . \quad . \quad . \quad . \quad . \quad . \quad . \quad (65$$

Therefore the ratio,

$$\frac{\text{Net cold feed water}}{\text{Total steam}} = \frac{\beta_1 \beta_2}{1 + \beta_1 + \beta_2} = \beta \text{ say} \quad . \quad . \quad . \quad (66)$$

Let $v_8$ be the final velocity of the delivery jet; then, if the steam for both forcer and lifter is taken from the boiler, we shall have the total momentum of the two steam jets equal to the momentum of the final delivery jet, or

$$v_2 = v_8 (1 + \beta),$$

and $$v_8 = \frac{v_2}{1 + \dfrac{\beta_1 . \beta_2}{1 + \beta_1 + \beta_2}} = \frac{v_2 (1 + \beta_1 + \beta_2)}{(1 + \beta_1)(1 + \beta_2)} . \quad . \quad . \quad (67)$$

This equation is only an approximation, as the pressure $P_8$ in the delivery jet has been neglected. From experiments on compound injectors, this has been found to be small.

We may find the temperature $t_8$ of the delivery jet thus—

$$1 + \beta_1 = \frac{1114 + {\cdot}3\, t_1 - t_5}{t - t_5}$$

also $$1 + \beta_2 = \frac{1114 + {\cdot}3\, t_1 - t_3}{t_8 - t_3}$$

therefore $$t_8 = t_5 + \frac{1114 + {\cdot}3\, t_1 - t_5}{1 + \beta_1} + \frac{1114 + {\cdot}3\, t_1 - t_3}{1 + \beta_2}$$

$$= \frac{t_5 (\beta_1 \beta_2 + \beta_2 - 1) + (1114 + {\cdot}3\, t_1)(1 + 2\beta_1)}{(1 + \beta_1)(1 + \beta_2)} \quad . \quad . \quad (68)$$

The compound injector is most advantageous when the "lifter" portion is an exhaust steam injector. The feed water is thereby heated to the same temperature as it would be with an ordinary single exhaust injector, and at the same time the "forcer" drives it into the boiler at any steam pressure now in general use, while the temperature is still further raised by the aid of the live steam.

The principal feature of the compound injector is the high temperature to which it will raise the feed water; so that in calculations with regard to this apparatus we may take approximate values for the velocity of the exhaust steam and of the delivery jet.

If we make $t_1 = 212$ deg. in the equation immediately following (67), we have—

$$1 + \beta_1 = \frac{1178 - t_5}{t_3 - t_5}$$

and

$$t_8 = t_5 + \frac{1178 - t_5}{1 + \beta_1} + \frac{(1114 + {\cdot}3 t_1)(1 + \beta_1) - \beta_1 t_5 - 1178}{(1 + \beta_1)(1 + \beta_2)}.$$

The gain of heat by the feed water due to the exhaust injector may be found thus: There is 1 lb. of exhaust steam for every $\beta_1$ lb. of feed water. This pound of steam contains $(1178 - t_5)$ thermal units, reckoned above $t_5$ deg, all of which would have been wasted, but by the aid of the injector are returned to the boiler. This, then, is evidently the net gain (*or saving*) of heat per $\beta_1$ lb. of cold feed water, and hence the saving per pound of water is—

$$\frac{1178 - t_5}{\beta_1} \text{ thermal units.}$$

Each of these pounds of water require

$$1082 + {\cdot}3 t_1 - (t_5 - 32) \text{ thermal units}$$

to evaporate it in the boiler; therefore the saving in heat expended per cent

$$= \frac{(1178 - t_5)\,100}{1114 + {\cdot}3 t_1 - t_5)\beta_1} \quad . \quad . \quad . \quad . \quad . \quad . \quad . \quad . \quad (69)$$

In the same manner, every $\beta_1$ lb. of cold feed water have mixed with them 1 lb. of exhaust steam; hence out of $(1 + \beta_1)$ lb. of feed that enter the boiler, the save in water must be 1 lb., or

$$\frac{100}{1 + \beta_1} \text{ per cent} \quad . \quad . \quad . \quad . \quad . \quad (70)$$

We see from this that it is expedient to use as little feed water as possible, or, in other words, as much exhaust steam as possible. If we assume that $t_5 = 60$ deg., $t_1 = 368$ deg., corresponding to a pressure of 170 lb. per square inch absolute, and $\beta_1 = 7{\cdot}5$, corresponding to a temperature $t_3 = 192$ deg., we have from (69) and (70) the save in heat $\doteq 12{\cdot}75$ per cent, and the save in feed water $= 11{\cdot}75$ per cent.

# CHAPTER XIV.

## QUANTITY OF FEED PER HOUR WITH EXHAUST INJECTOR.

WITH any exhaust injector we may easily find approximately the amount of feed water supplied to the boiler per hour. Let Q = this quantity in pounds, then $\frac{Q}{3600}$ lb. are delivered per second; and if D = the diameter of the delivery cone orifice in inches, its area must be

$$\frac{\pi}{4} \times \frac{D^2}{144} \text{ square feet,}$$

and the amount of steam and water fed to the boiler per second = area × velocity × density,

$$= \frac{\pi}{2} \times \frac{D^2}{144} \times v_3 \times \rho$$

Substitute $v_3$ by its value given in the equation

$$v_3 = \frac{v_2}{1+\beta}.$$

and assume that $v_2 = 1{,}200$ ft. per second; also that $\beta = 7$ (this is nearly its minimum value). We then find that the total delivery per second

$$= \frac{\pi}{4} \times \frac{D^2}{144} \times 62{\cdot}5 \times \frac{1200}{8} = 51 \text{ D}^2.$$

Of this quantity of water and steam delivered every $(1+\beta)$ lb. are made up of $\beta$ lb. of water and 1 lb. condensed steam; hence of the above quantity $\frac{\beta}{1+\beta} \times 51 \text{ D}^2$ lb. are feed water, and $\frac{1}{1+\beta} \times 51 \text{ D}^2$ lb. are condensed steam. Then

$$\frac{7}{1+7} \times 51 \text{ D}^2 = \frac{Q}{3600}$$

or

$$Q = 160650 \text{ D}^2 \quad . \quad . \quad . \quad . \quad . \quad . \quad . \quad (71)$$

where D is measured in inches.

If D should be measured in millimetres, then

$$Q = 160650 \times \frac{D^2}{25{\cdot}4^2} \quad . \quad . \quad . \quad . \quad . \quad (72)$$

The weight of steam used per hour in pounds

$$= \frac{Q}{\beta} = \frac{Q}{7} \text{ lb.}$$

Volume of steam used per minute

$$= \frac{Q}{7} \times \frac{\text{volume of 1 lb. at 212 deg.}}{60}$$

$$= \frac{Q \times 26{\cdot}37}{420} = {\cdot}063 . Q \text{ cubic feet} \quad . \quad . \quad (73)$$

This being the volume of steam used by the injector, the supply from the exhaust must never be less than this amount. So that the injector shall not have its action impaired for want of steam, makers recommend that for good working there should be at least one and one-third times this amount, and when an early cut-off happens, such as with a locomotive or a variable expansion engine, the minimum volume of steam should be twice that given by the above equation.

The rule given by Messrs. Holden and Brooke to determine the maximum size of injector that can be used is this:

Find the volume swept out by the piston per minute in cubic feet, and multiply it by the fraction three-quarters; then look in the subjoined table, and see what number of

TABLE VII.

| Size of Injector. | 2½ | 3 | 3½ | 4 | 4½ | 5 | 6 | 7 |
|---|---|---|---|---|---|---|---|---|
| Minimum cubic feet of steam required per minute...... | 43 | 60 | 80 | 100 | 126 | 156 | 225 | 306 |
| Maximum of water delivered in pounds per hour....... | 95 | 150 | 200 | 270 | 340 | 420 | 600 | 830 |

TABLE VII.—*Continued.*

| Size of Injector. | 8 | 9 | 10 | 11 | 12 | 13 | 14 | 20 |
|---|---|---|---|---|---|---|---|---|
| Minimum cubic feet of steam required per minute...... | 400 | 506 | 625 | 756 | 900 | 1056 | 1225 | 2500 |
| Maximum of water delivered in pounds per hour....... | 1080 | 1370 | 1700 | 2050 | 2450 | 2870 | 3330 | 6800 |

injector will require this amount. If the cut-off takes place very early, replace the fraction three-quarters by one-half.

The above gives the maximum amount of feed delivered by the minimum amount of steam. The data which really must determine the size of the injector is the size of the

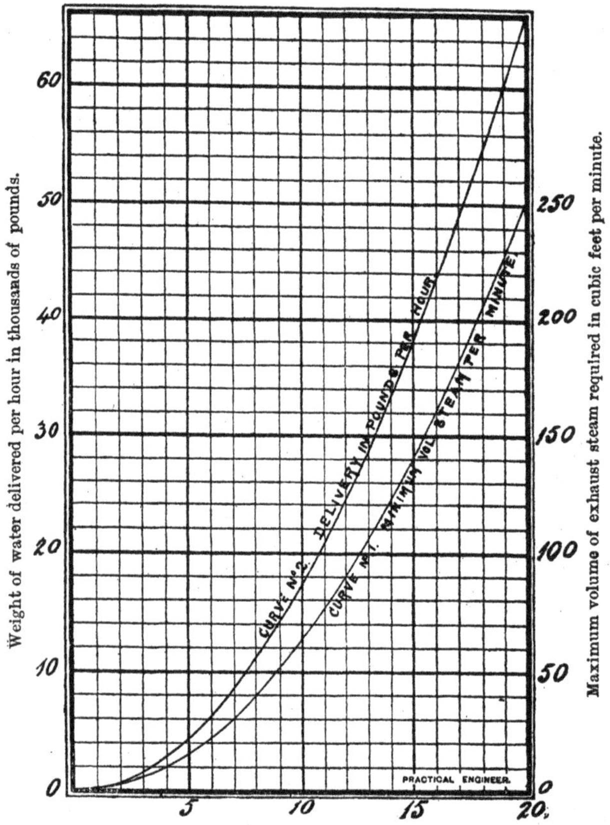


FIG. 30.—Number of Injector and diameter of delivery cone in millimetres.

boiler to be fed, and an exhaust injector should not be any larger than is necessary; otherwise the boiler will get filled too quickly, and the working must then be intermittent, and thus the maximum saving by the use of the exhaust steam will not be obtained. This is a point which should receive careful consideration when procuring an injector. It may appear strange that the above figures are given almost irrespective of the pressure of the steam when first admitted into the cylinder of the engine; but upon a little reflection it will be seen that it is not the almost instantaneous puff of exhaust steam that supplies the injector with motive power, but it is the exhaust pipe and cylinder full of steam at atmospheric pressure which is left behind after the puff has taken place, the steam being supplied continually by the movement of the piston in the cylinder. Thus it is that the exhaust injector may reduce the back pressure in the cylinder when it is slightly above that of the atmosphere. The result of Table VII. has been plotted in fig. 30, the abscissæ being the number of the injector—*i.e.*, the diameter of the delivery cone in millimetres, and the ordinates (1) the minimum volume of steam required per minute, (2) maximum weight of water that can be delivered per hour in pounds.

---

## CHAPTER XV.

### Exhaust and Compound Injectors (Descriptive).

The principal makers of exhaust injectors in this country are the Exhaust Injector Company and Messrs. Holden and Brooke; and it is chiefly with apparatus manufactured by these two well-known Manchester firms that we shall deal in the following pages. Messrs. Schäffer and Budenberg make a very large number of exhaust injectors, as they hold the foreign rights of the flap-nozzle patent, but they do not sell them in the United Kingdom. The construction of their injectors is similar to those made by the Exhaust Injector Company.

From the nature of the supply of exhaust steam from a direct-acting engine, it is evident that the *re-starting* capabilities of an exhaust instrument must be of the most sensitive description. This feature is the very essence of successful working, and without which the exhaust apparatus remains an inert piece of material only.

A longitudinal section of an injector made by the first-mentioned firm above is shown (fig. 31) as fitted to feed a stationary boiler. In appearance it resembles the flap-nozzle re-starting injector, with the exception that the steam nozzle is very large, the combining cone of extra length, and a conical spike is fixed axially in the steam cone. The excessive size of the steam orifice has already been predicted in the mathematical passages of the previous pages. The spike, it is said, tends to assist in the concentrating of the jet of exhaust steam. It will be noticed that the overflow pipe is of peculiar construction, having a vertical wall across the main portion; at the same time the end of the pipe remote from the injector dips below the surface of the water in a tank. The object is to prevent the inlet of air

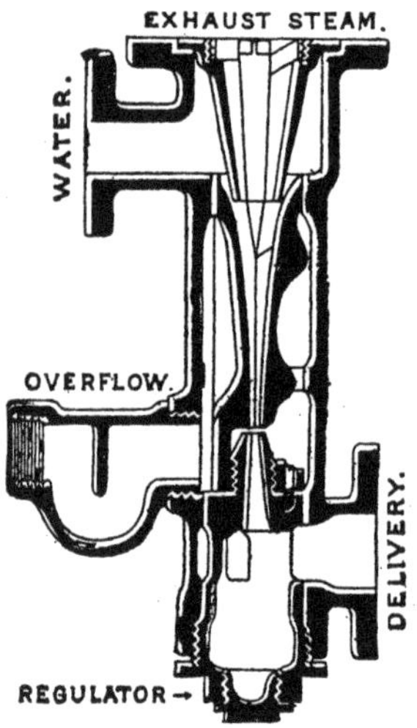


FIG. 31.

to the combining cone of the injector, which would immediately stop its action. This may be gathered from the previous mathematical treatment, where it was shown that the pressure in the combining cone was considerably less than that of the atmosphere.

The general arrangement of the injector is shown in fig. 32. The junction with the exhaust pipe is perfectly plain, and free from projections, which tend to the increase of back pressure in the cylinder, while the water must be supplied to the apparatus from above the injector. A small throttle valve, having a hand-wheel attached, is placed in the steam pipe just above the injector, to regulate the supply of exhaust steam. It has been previously mentioned that to secure the maximum efficiency of any injector the steam must be as dry as possible. The steam coming from the exhaust port of an engine often contains a considerable and variable amount of moisture, none of which should enter the injector. To secure this end the junction with the exhaust pipe should never be made in such a position that condensed steam can gravitate to the injector. Examples of general practice may be gathered from the various diagrams

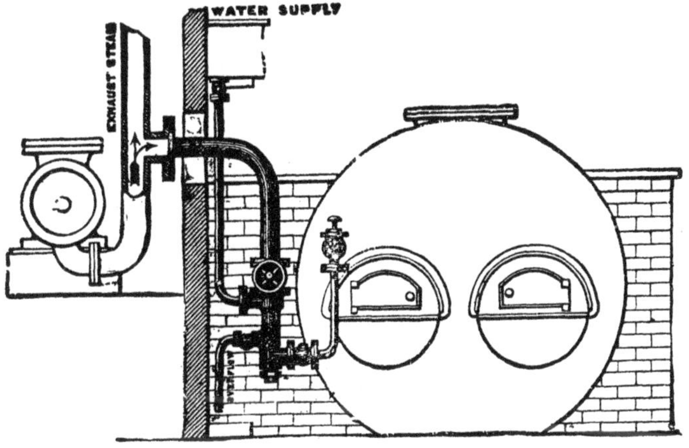


FIG. 32.

which follow. The above remark regarding the junction with the exhaust pipe should not be lightly passed over, as, if neglected, the efficiency of the apparatus may be impaired.

The supply of water coming into the combining cone can be regulated by turning the nut at the delivery end (fig. 31), thus making the combining cone advance towards or recede from the steam cone. It must be remembered that, should the injector be placed with its axis horizontal, the flap must *always open upwards* for best working. It can be made to open in a horizontal direction, and it has worked when opening downwards. If it is necessary to work the injector when the engine is standing, a small pipe connects the steam space in the boiler with the steam space in the injector just below the wheel valve (fig. 32). In this way the boiler steam may be wire-drawn down to the pressure of the atmosphere, and used exactly as the exhaust steam from the engine. Should the boiler pressure be between 70 lb. and 105 lb. per square inch, the makers recommend the use of the supplementary live steam nozzle (fig. 33), whereby the jet is assisted by live steam from the boiler *after* it has been mixed with the exhaust steam. The figure explains itself.

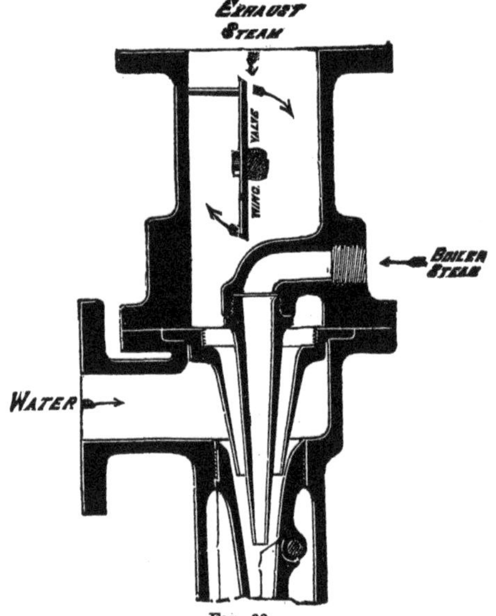


Fig. 33.

The exhaust injector of Messrs. Holden and Brooke is of a somewhat different type to the one just described, but works within about the same limiting conditions. The illustrations which the author has are all in connection with boilers of a higher pressure than about 80 lb. per square inch; which will be fully dealt with when we come to "compound injectors"; but their single injectors are coupled up and managed in a manner almost identical with that just described.

# CHAPTER XVI.

## COMPOUND INJECTORS.

THE ordinary compound live-steam injector is not used to any great extent in this country, though on the Continent and in America it has gained more favour perhaps than the single instrument.

A longitudinal section of Messrs. Körting Brothers' non-lifting injector is shown (fig. 34). It is essentially two single injectors; the one feeding the other, and the second feeding the boiler. The object of the former is to supply

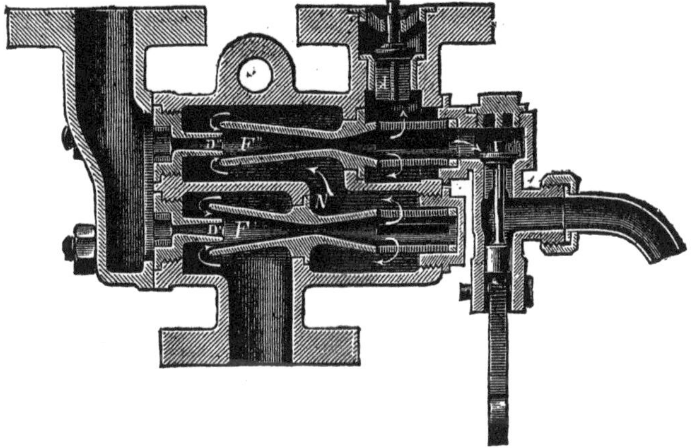


FIG. 34.

the latter with feed water under pressure, so that the final temperature shall be above 212 deg. The steam cone of the lifter (lower portion) is considerably smaller than that of the forcer. There is no overflow opening in either of the cones; but there is a valve in the delivery pipe which is opened at starting, and which acts as an overflow valve for the time being.

The delivery water from the lifter follows the direction of the arrows through the chamber N, and thence into the second combining cone F. In this injector all the cones can be removed without breaking any pipe connections. The apparatus must be placed at least a foot below the lowest water line, but will then take feed-water up to a temperature of 155 deg. It will be seen from the figure that no feed-water valve is required,

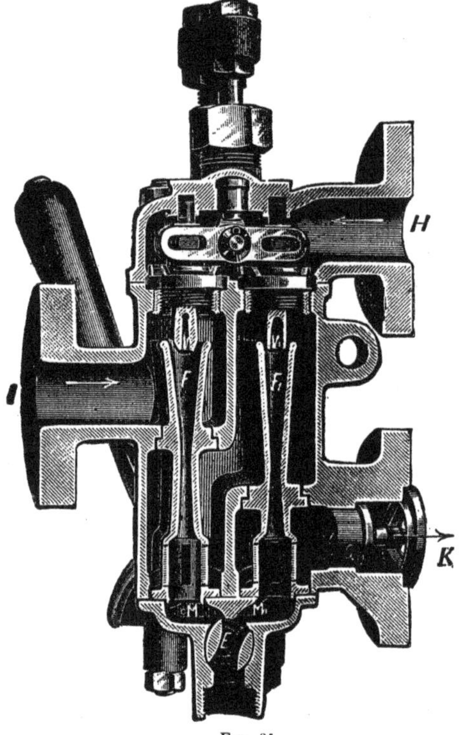


FIG. 35.

the starting valve preventing any waste. This closes automatically.

A *lifting* compound injector by the same makers is shown (fig. 35), in which the general idea is much the same as in the last example. A pair of valves are added to the steam cones, both being operated simultaneously by one handle.

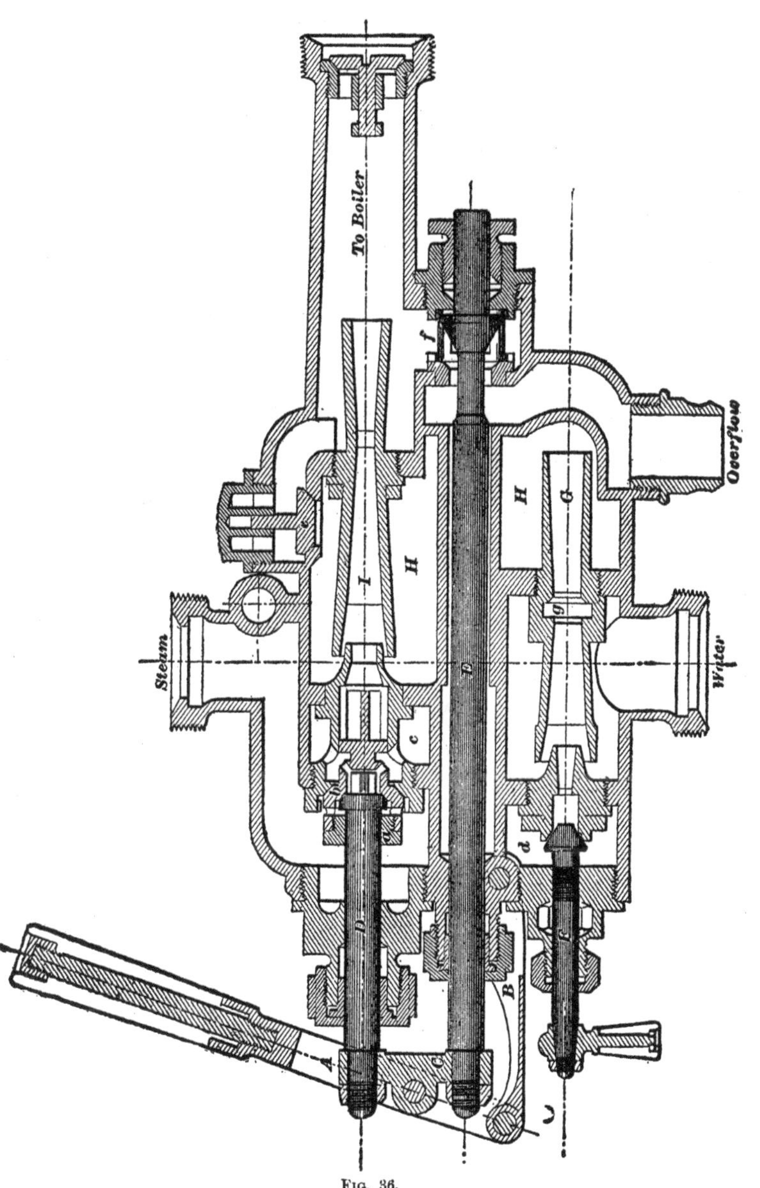


Fig. 36.

open
shut
D
Practical Engineer
WATER
STEAM
B
C
A
OVERFLOW
TO BOILER
A

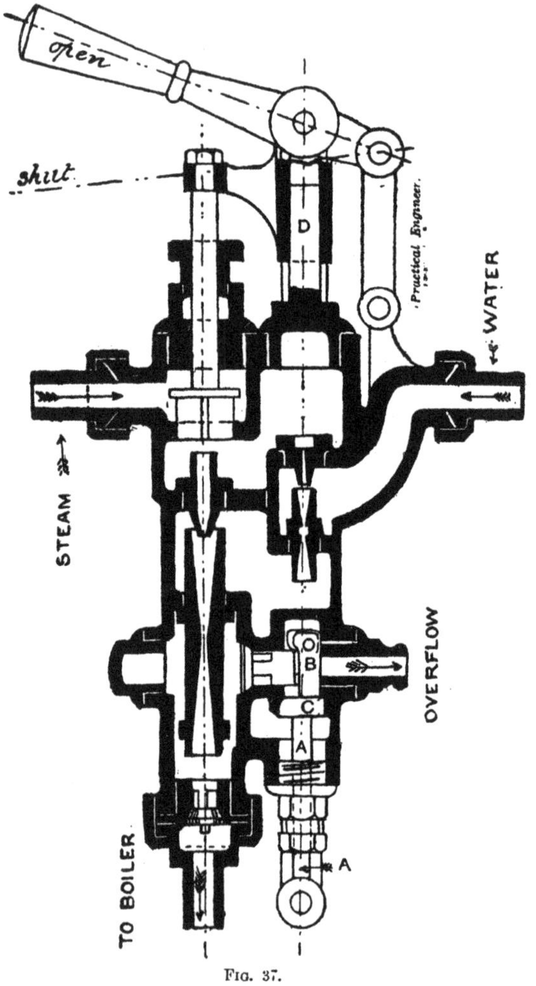

Fig. 37.

The delivery chambers of both forcer and lifter communicate with the overflow pipe by the movable cock E and the channels M M. To start the injector the handle is moved through a part only of its travel, thus slightly opening the steam valves, and at the same time opening both delivery chambers to the overflow pipe. This will exhaust the feed-water pipe and lift the feed to the injector, after which the handle is moved right over, fully opening both steam valves and completely closing the overflow, when the injector begins to feed the boiler. The injectors may be placed in any convenient position, and will lift over 20 ft. with cold feed water. The maximum temperature of the feed water with an overhead supply is 150 deg.

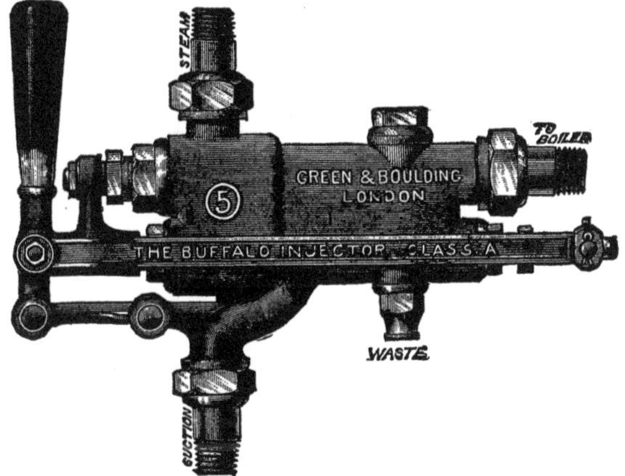


FIG. 38.

It is not necessary to regulate the steam or water supply for variation of steam pressure, as the quantity passing through the injector depends upon the lifter ; so that if by a decrease of pressure less water enters the lifter, there will be less for the forcer to drive into the boiler.

Some experiments on one of these injectors show that the amount of water delivered per hour varies according to the temperature of the feed, a fact which the author has previously pointed out. A non-lifting injector, No. 4, delivered 2,000 lb. of cold water, but only about 1,600 lb. of hot water, with steam at 60 lb. per square inch.

The same makers use these injectors in conjunction with feed-water heaters on the German State railways. The heaters are supplied with exhaust steam from the blast pipe. It appears that English locomotive engineers will not adopt anything of the kind, and the makers have practically dropped the matter.

A very good improved form of compound lifting injector is shown in section, fig. 36. The chief features of this apparatus are the automatic action of the overflow valve *e*, and the entire regulation of steam to both lifter and forcer by the single valve stem D. The little supplementary regulator F is only attached to locomotive injectors. There is also an attempt to

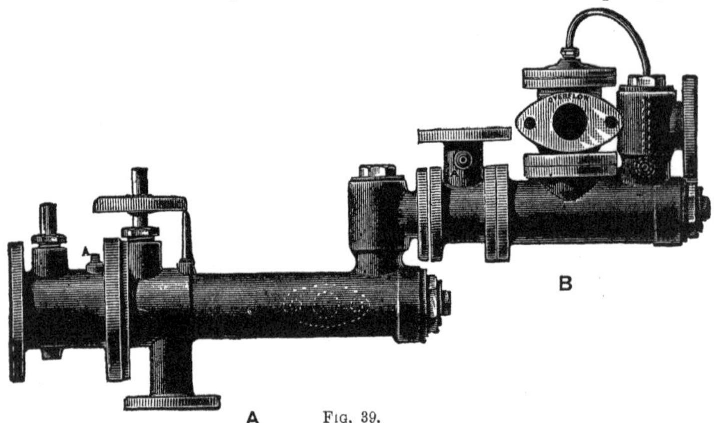


Fig. 39.

automatically regulate the supply of feed water delivered to the forcer by the lifter with the orifice *g* between the combining and delivery cones of the lifter. Should too much feed pass with the steam into the combining cone, the surplus will return to the feed chamber through the passage *g*; but should there not be sufficient feed entering with the steam, then more will be induced into the cone G through the aperture *g*.

To start the injector, the handle is slightly lifted, thereby raising the collar valve on the end of the stem D from its seating, and allowing a small amount of steam to pass through the main valve *b* into the chamber *c*, and thence into *d*, from which it flows into the lifter and raises the water to the injector. On the arrival of the water, which pushes back the overflow valve *e*, thence through *f* and out at the overflow pipe, the handle is slowly raised to its extreme position, opening the main steam valve *b* and gradually closing the overflow valve *f*, after which the injector is at work. The above injector is of American design, the makers being Messrs. Hayden and Derby, of New York.

The injector shown in section, fig. 37, and elevation, fig. 38, is called the "Buffalo" injector, and is the specialty of Messrs. Green and Boulding, of London. The design has a decidedly American appearance, and is very compact and handy, while all the moving parts can be examined easily while the injector is under steam. Steam is admitted to both forcer and lifter at the same time by raising the valve

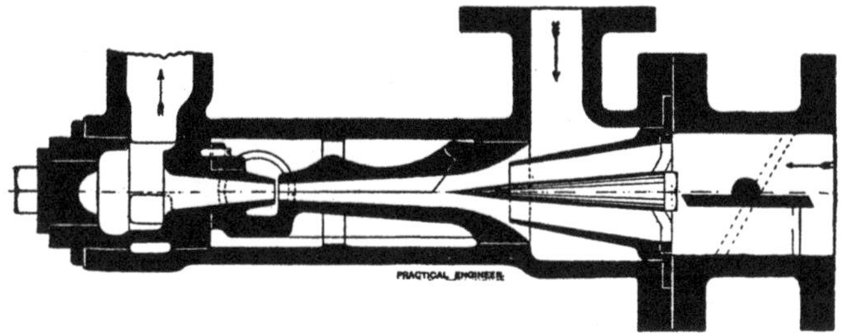


FIG. 40.

at the end of the steam pipe. The overflow valve, controlling the overflow orifice of the forcer, is operated by the spindle A, and the cam arrangement at B. When the hand lever is down in the position "shut," the spindle A is also depressed into its lowest position, by means of the connecting link, shown more plainly in the elevation (fig. 38). In this position the cam at B has forced the overflow valve open, and steam and water can escape into the atmosphere from the forcer, and from the lifter past the little piston C into the atmosphere. When the water has arrived, and condenses the steam, a pressure is produced in the discharge chamber of the forcer, which tends to close the overflow valve; at the same time the pressure of the water in the lifter discharge chamber forces the little piston C into the position shown, closing the overflow from the lifter, and assisting in opening the steam valve. The spindle D acts as a guide to the crosshead which operates the steam valve. In ordinary locomotive practice the

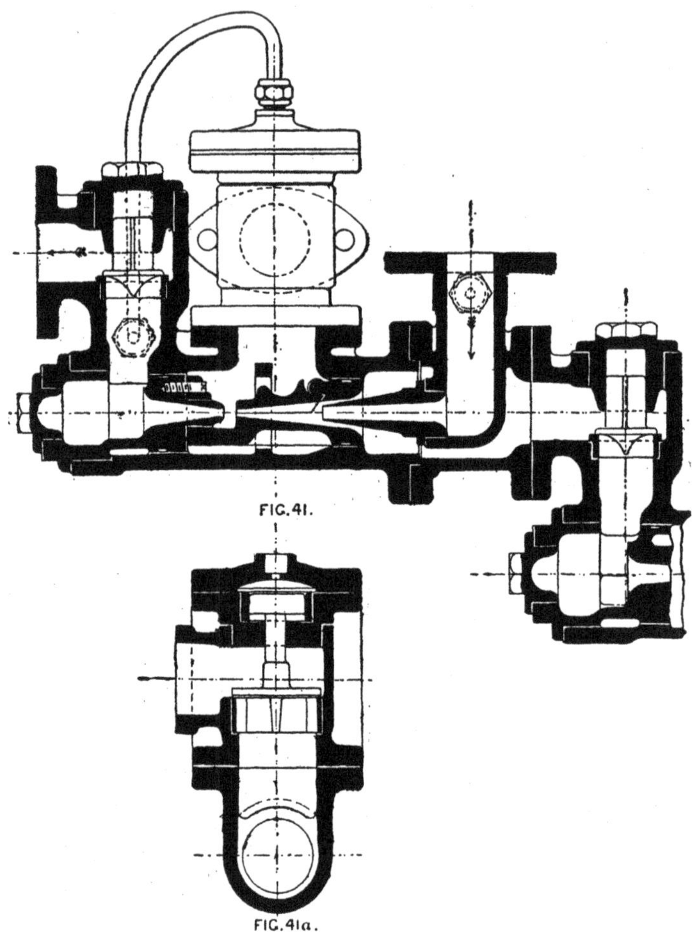

FIG. 41.

FIG. 41a.

prevailing pressures range between 100 lb. and 200 lb. per

square inch; and hence a special device is necessary beyond that yet described to make the exhaust injector do duty as a boiler feeder. This is very ably and simply brought about by using a compound injector, of which the

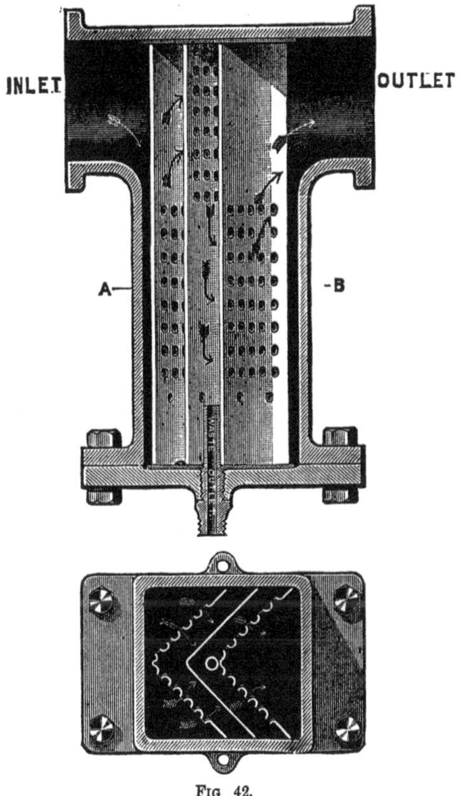


FIG 42.

lifter portion is an exhaust injector and the forcer an ordinary live steam re-starting injector. A side elevation of one of these instruments is given in fig. 39. The exhaust steam enters at the left-hand end, and the cold feed water at the lowest flange. The portion A is the exhaust injector,

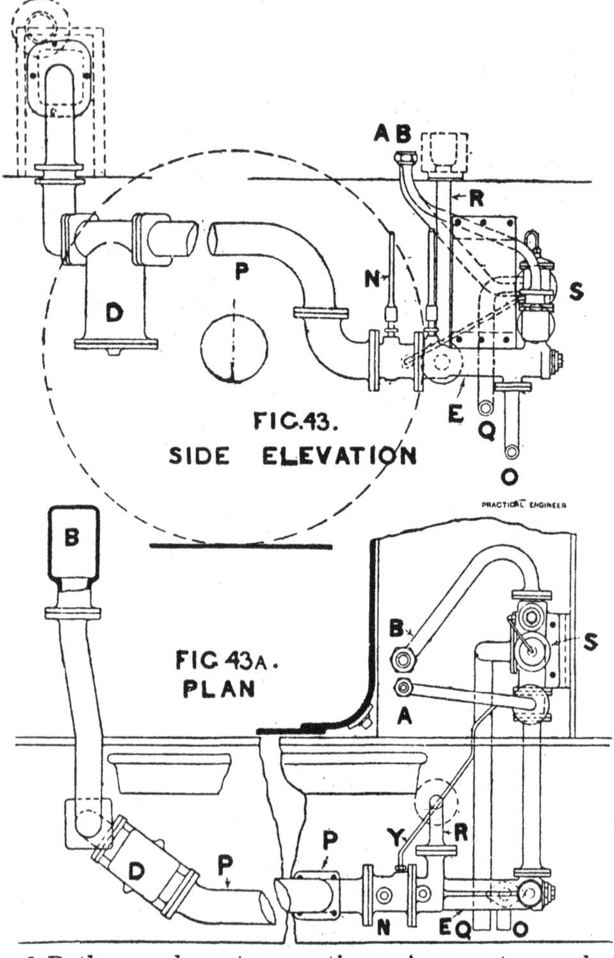


and B the supplementary portion. A non-return valve separates the two portions.

A casting crowns the overflow flange. This contains an overflow valve, which is maintained upon its seating, while the injector is at work, by the pressure of the delivery jet being communicated to a little piston on the top of the valve spindle. The necessity for this is evident with a compound apparatus. There is a small pipe seating A close to the water regulator, which is connected by a small pipe to another pipe seating A on the live steam supply pipe. By the use of this pipe the injector may be worked when the engine is standing.

A section of the exhaust portion A is shown, fig. 40, and a section of the supplementary portion, fig. 41. For the different views of this injector we are indebted to the makers,

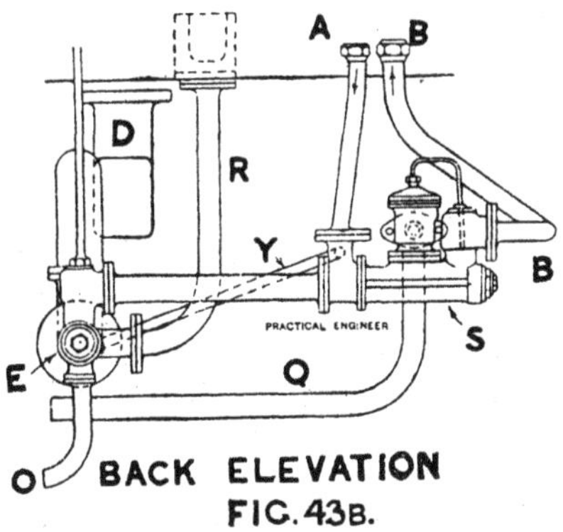


BACK ELEVATION
FIG. 43B.

the Patent Exhaust Injector Company, and to their manager, Mr. A. Slater Savill, M.I.M.E. The shut-off valve, fig. 39, is used to prevent any exhaust steam entering the injector while it is not being used, making it very hot. The overflow pipe, fig. 40, has a non-return valve in it, so that the atmosphere and its pressure may be kept out of the combining tube.

The construction of the supplementary portion or "forcer" is well shown in the section, fig. 41, and the supplementary overflow valve, fig. 41 *a* (page 108). When the injector is starting, the pressure above the little piston of this valve is the same per square inch as that below the valve itself, namely, that of the steam jet in the combining cone; and as the area of the valve is greater than the area of the piston, the valve necessarily lifts. Thus the supplementary portion is quite as much a lifting and re-starting apparatus as a single injector with the flap nozzle. To start this injector, the water is turned on first, and then the live steam to the supplementary portion, after which the injector immediately starts to work. By turning on the live steam, the air in the exhaust portion is exhausted, and the exhaust steam from the blast pipe is induced to flow to the injector instantly.

To prevent any grease or deposit from the cylinders entering the boiler, and thus reduce its evaporative efficiency,

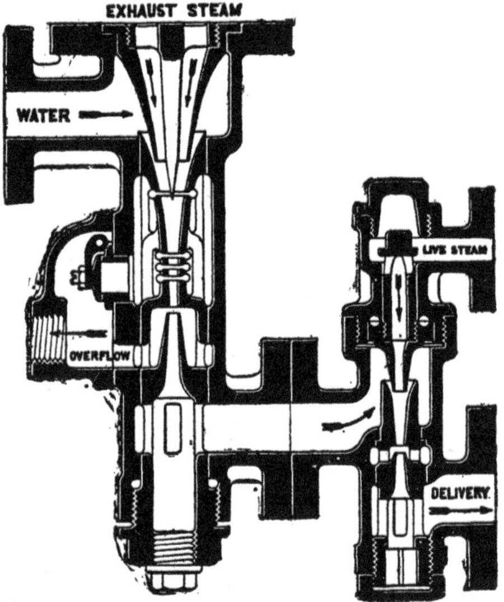


Fig. 44.

as well as endanger the life of the firebox, the exhaust steam, as it comes from the blast pipe, passes through a rectangular box, containing perforated diaphragms, fig. 42, and called a "grease separator." This is the invention of Mr. Savill, and is used on all locomotives with his firm's exhaust injectors.

It must be remembered that it is only used on locomotives, where a strong blast is necessary for the production of a good draught. The general arrangement of the whole of the pipes, connections, &c., is shown in side elevation, fig.

43; in plan, fig. 43A; and in end elevation, fig. 43B, as fitted on about 30 mineral tank engines on the Barry Railway, South Wales. The blast pipe is shown at B in section, and the grease separator at D, from which the main exhaust pipe conveys the exhaust steam to the injector through a cast-iron pipe P, which runs nearly the whole length of the engine underneath the side footplates. N is the exhaust regulator, and R the feed-water supply pipe. The exhaust overflow pipe is shown at O, and the supplementary at Q, both being brought into a position so that their orifices may be seen from the cab by the fireman. The live steam pipe is given at A, and the delivery pipe at B. The supplementary portion S is situated at right angles to the exhaust portion E, and immediately under the feet of the fireman. The

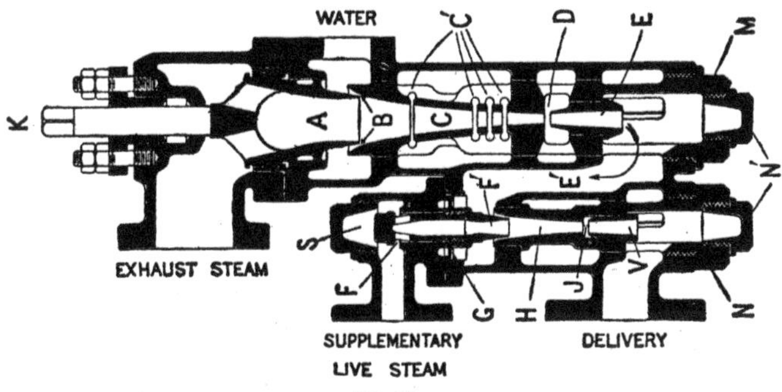


FIG. 45.

small pipe Y supplies live steam to the exhaust injector, so that it may be worked when the engine is stopped.

These injectors, which are fitted on one side only, have given great satisfaction to both drivers and the locomotive superintendent, Mr. J. H. Hosgood, M.I.M.E. The boilers steam much more freely than with the ordinary single live-steam apparatus, and during the three years they have been in constant use the average saving in coal consumption over similar sister engines, working the same traffic, is about 4 lb. of coal per train mile, or about 10 per cent. After the drivers got over their prejudice, they preferred engines fitted with the exhaust apparatus, and found no more difficulty in using them than those worked by live steam only. The author considers that, if injectors can be worked upon the Barry railway engines, they can be adopted without fear almost anywhere, as the whole district traversed by the railway is of the very worst description for locomotives, or, in fact, for any boiler, the water containing excessive quantities of lime and magnesia in solution. The working of the apparatus above described is very excellent, and leaves little to be desired.

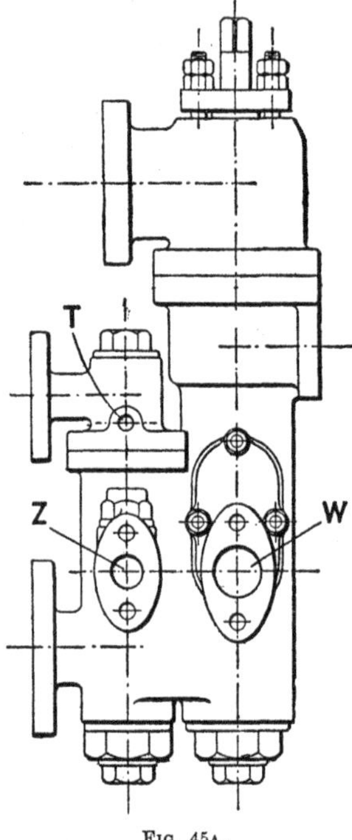


Fig. 45a.

The accompanying contour shows the gradients of the main line over which the Barry engines run, the nature of which can be gathered from a glance at the figure. The dimensions of the engines fitted with the exhaust injector are identical with those fitted only with the re-starting instrument. They are chiefly—diameter of cylinders, 18 in.; diameter of coupled driving wheels, 4 ft. 3 in.; steam pressure by the gauge, 150 lb.; weight of water in side tanks, 1,400 gallons.; weight of coal carried, 35 cwt.; total weight of engine in running condition, 49 tons; average speed, 17·5 miles per hour; heating surface, 1,288 square feet; grate area, 21 square feet. The average distance run by each engine per day is about 90 miles, and the coal used

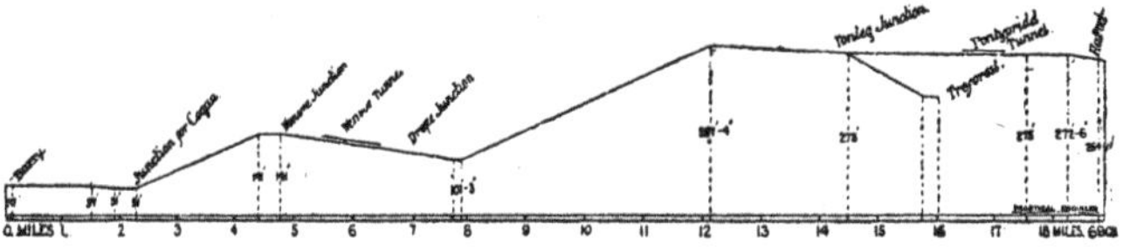

Contour of Barry Railway Main Line.

comes from the Great Western Colliery. There are 32 engines fitted with the exhaust injector, and 24 similar sister engines with the re-starting injector only. A comparison of the two average performances is here given :—

| Type of injector. | Average load up in empty 10 ton wagons. | Average load down in fully loaded 10 ton wagons. | Total load up including engine and van in tons. | Total load down including engine and van in tons. | Average load up and down in tons. | Pounds of coal per train mile. |
|---|---|---|---|---|---|---|
| Restarting .... | 42 | 38 | 266 | 626 | 446 | 43·3 |
| Exhaust ...... | 40 | 45 | 256 | 731 | 494 | 39·3 |

The above indicates a net saving of coal for the exhaust injector over the live-steam apparatus to the amount of 4 lb. per train mile, or 9·2 per cent, and if we take into consideration the extra average load, it reaches about 10 per cent.

The above exhaust injectors have been fitted up and in daily use since February, 1890, or something over three years. Both the live-steam and exhaust apparatus require to be examined and cleaned at Barry on the average every three months. The shortness of this period is, of course, caused by the bad state of the water. Similar injectors working in other parts have been at work for years without being touched.

With the exhaust injectors above described the feed water can be easily taken when at a temperature of 90 deg.

Fah., with 200 lb. boiler pressure. In ordinary working,

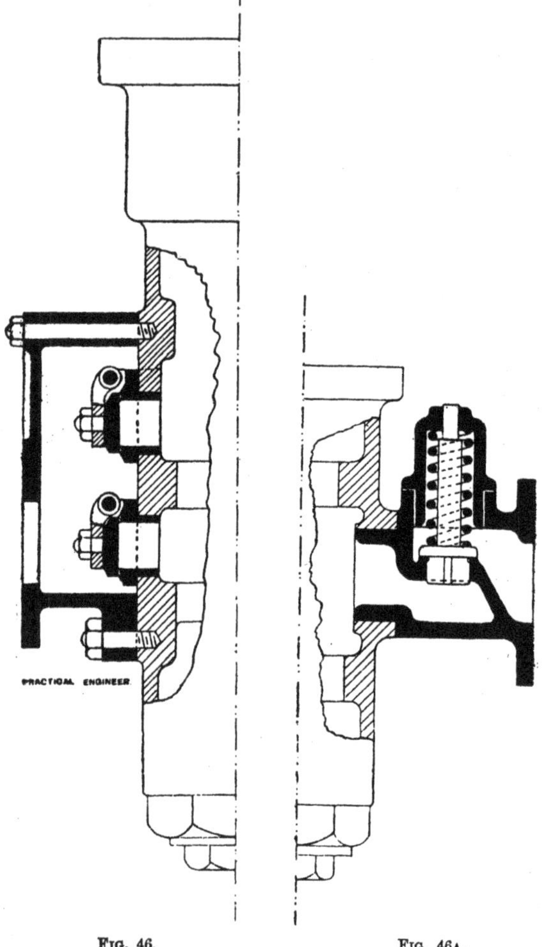


FIG. 46. FIG. 46A.

with boiler pressure at 150 lb., even in the coldest of winter weather, the feed water leaves the exhaust portion at a temperature from 180 deg. to 198 deg., and the supplementary then delivers the same water into the boiler at a temperature of 280 deg. Fah. It is at once evident that this class of boiler feeder is at the same time a most excellent feed heater.

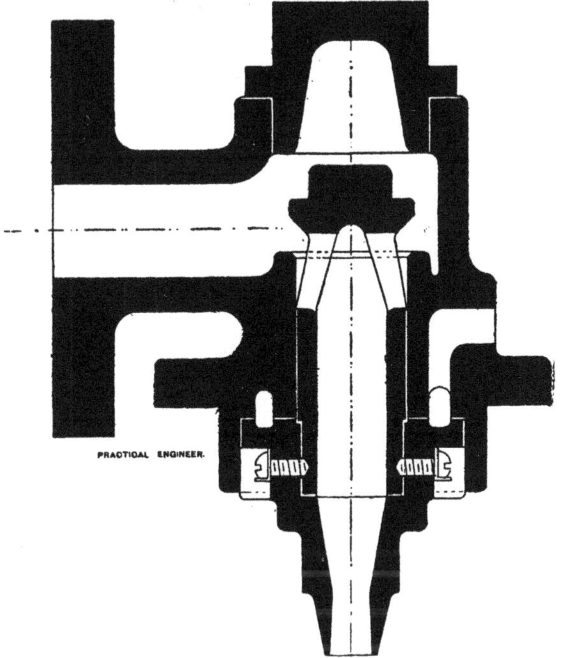


FIG. 47.

Another form of exhaust injector for feeding boilers of high steam pressure is given in fig. 44, and is the patent of Mr. R. G. Brooke, of the firm of Holden and Brooke, of Salford. The exhaust (left hand) portion is their ordinary form of exhaust injector, which they use for all purposes when the pressure of the boiler to be fed does not exceed 75 lb. or 80 lb. per square inch. It much resembles the same firm's re-starting influx live-steam injector, and differs from it in having three extra transverse slits in the combining cone near the overflow orifice; the four slits communicating with a separate chamber, whose outlet is covered by an ordinary flap valve. These slits and flap valve act in the same manner as the flap nozzle just previously described, and render the action perfectly automatic. The amount of cold feed water can be regulated by turning the nut at the bottom of the delivery chamber, the annular water orifice being increased or

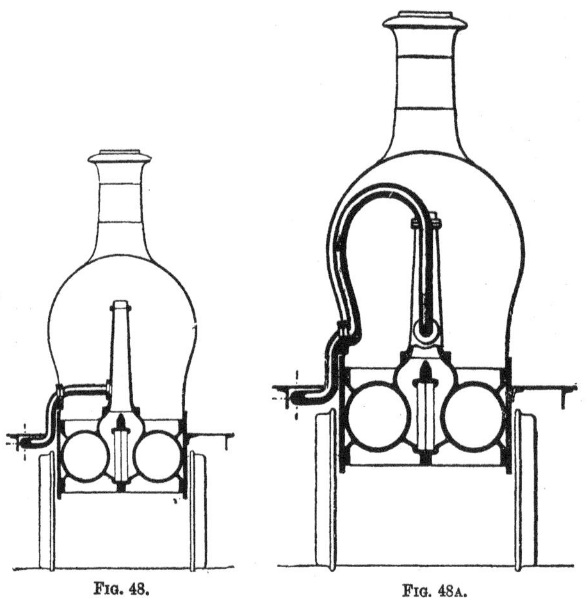

Fig. 48. Fig. 48A.

decreased thereby, as in the flap-nozzle instrument. The combining cone is strengthened and maintained in true shape by longitudinal ribs, shown in the figure in elevation. The chamber surrounding the flap valve is bolted to the main casing, and can be removed without affecting the remainder of the apparatus. The supplementary, or right-hand portion, is bolted to the delivery flange of the exhaust injector, and in this way may be fitted to any existing injector without much alteration. It consists of an ordinary simple live-steam injector, with a spring-controlled overflow valve, and an exceedingly ingenious and beautiful automatic valve for admitting the boiler steam. An enlarged detail sketch of this valve in section is shown in fig. 47. It will be seen that the steam cone is movable, and

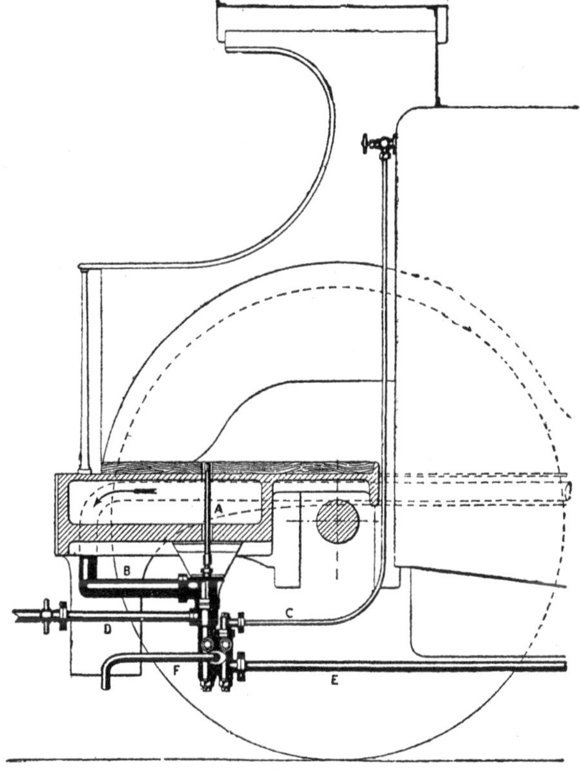


FIG. 49.

slides axially; also that it is capped by an ordinary mitre valve, which fits steam-tight upon its seating in the main casting. The lower portion of the cone is cast with a flange, which is accurately machined to slide in a corresponding cylindrical aperture in the main casting. The flat face of the flange is ground to a steam-tight fit with emery powder, so that when in the uplifted position, shown fig. 47, no steam or water can leak out into the atmosphere through the opening to the right. The action of this automatic valve is at once seen from an inspection of fig. 44. The water is turned on to the injector, and the

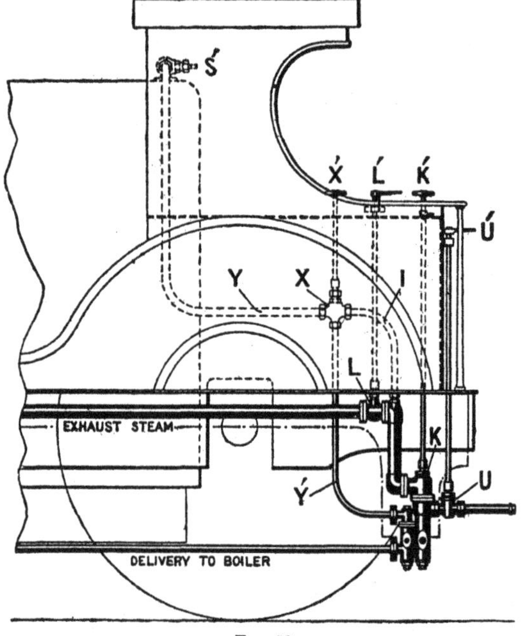


Fig. 50.

exhaust steam delivers it under pressure to the supplementary combining cone. When the pressure there is great enough, it will lift the steam cone, by virtue of the large area of the flange piston, and thus open the steam valve, admitting steam to force the feed water into the boiler. By this arrangement only one movement is necessary to start the injector, namely, that of the cold feed water supply valve.

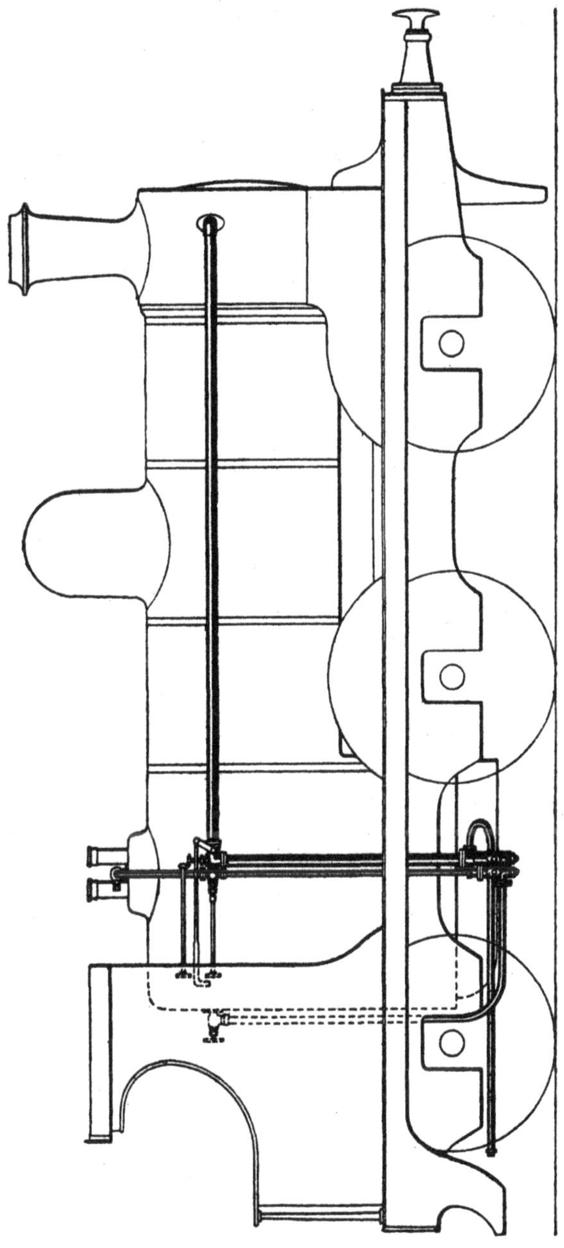

Fig. 51.

A very neat and compact arrangement of this injector, as used on locomotives, is shown in section, fig. 45, and elevation, fig. 45A. The outside casing is formed with a single casting, the axes of the two portions being parallel. Instead of regulating the water by a nut at the delivery end, as in the previous example, it is done by a screw movement of the exhaust steam cone, which can be actuated by a key on the spindle K. This allows of regulation being accomplished without leaving the footplate, though it is seldom required except after cleaning or repairs. The overflow valves of the exhaust portion are shown in detail in fig. 46, and that on the supplementary portion in 46A. The exhaust steam flange on the injector is connected to the blast pipe by a pipe, the smokebox end of which is shown in fig. 48. If this method of connection should interfere in any way with the cleaning of the tubes, the pipe may be bent round so as to avoid this, as in fig. 48A, the joint being made in front of the blast pipe instead of on the side. This is probably the better method, as there is a tendency to superheat the exhaust steam on its way to the injector by the waste furnace gases.

When the injector is placed immediately below the footplate, the levers and pipes may be arranged as in fig. 49. This is exceedingly simple, and one of the first brought out by this firm. The exhaust steam pipe B runs along under the side footplate, quite out of sight, and enters the injector at the top. The spindle A is for operating the cold-water regulator, and, as this is seldom done, its key is removed to the tool box. Live steam reaches the injector by the small pipe C, while E is the delivery pipe.

An improved arrangement is shown in fig. 50, in which the appartus may be used either as an exhaust or live steam injector. This is accomplished by attaching the four-way junction and valve X to the small live-steam pipe Y leading to the supplementary portion, and connecting X with the exhaust steam pipe just behind the wing valve L by the small pipe I. The *modus operandi* is as follows: The wing valve L closes the exhaust pipe, after which the valve in X is opened by the hand wheel $X^1$, admitting live steam, wiredrawn down to atmospheric pressure into the exhaust portion. By the wing valve L the amount of exhaust steam may be easily regulated, though this is not a general occurrence when the engine is doing the same work for a length of time. Of course the wing valve must be opened to start the injector, but it is customary for drivers to mark the position in which the injector works best, and to turn the handle gradually on to the mark when starting.

When the blast is very strong it is found that the wing valve should be only partially open, otherwise the excessive blast may throw the injector off and blow the water back into the tender.

Fig. 51 shows a method of fitting the injector in front of the cab, in which all the different parts are easily accessible for examination. The chief feature in this figure is the patent arrangement of combined fittings, situated on a level with the firebox crown, and shown in detail in fig. 52.

The object sought is to be able to regulate the exhaust steam and the water, and to admit live steam to the exhaust

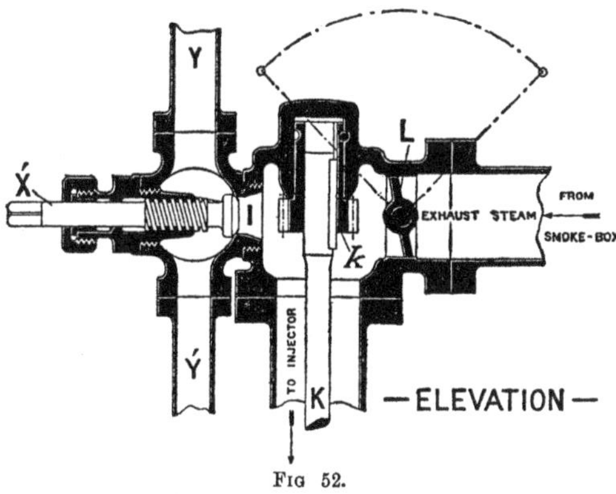


Fig 52.

portion, to work the injector while the engine is standing, all with a single fitting. The live-steam pipe Y, fig. 52, has an aperture I, closed by the valve X, to admit live steam to the exhaust pipe. The spindle K, operated by a worm wheel and worm (not shown), runs down to the water regulator in the exhaust injector, and seldom requires to be touched. The only movement required in this combined fitting to start the injector is that of the wing valve L.

The amount of water delivered per hour is approximately the same as that delivered by a similarly-sized single exhaust injector ; but the temperature of the delivery is raised about 60 deg. or 70 deg. above that with an exhaust injector alone. Generally speaking, the temperature of delivery of the exhaust portion ranges between 180 deg. and 200 deg., and the temperature of the feed as it enters the boiler about 260 deg. It is thus evident that a considerable saving of coal and water ought to ensue from the use of exhaust steam in the "lifter," and at the same time the feed water enters the boiler at a temperature of from 80 deg. to 100 deg. higher than when fed by a single live-steam injector, and the stresses in the boiler shell, set up by the cooling effect of the comparatively cold feed water, avoided to a great extent. These injectors take feed water up to 80 deg. Fah., in which case the supply tank should be above the combining cone. For hot climates a special injector is designed to take feed up to 90 deg.

A few remarks with respect to the method of working and maintaining these injectors may not be out of place here. The live-steam supply to the supplementary portion should always be maintained at full boiler pressure, and never wire-drawn; hence the steam stop valve must always be left full open. The supply of exhaust steam may vary under different conditions of load; hence the exhaust steam wing valve may want altering in position under various circumstances. The exhaust steam should not exceed the atmospheric pressure when it reaches the injector; hence when working entirely with live steam, that which goes to the exhaust portion must be wire-drawn considerably. The condition of the steam in the injector may be judged from the appearance of the overflow. In the figures a grease separator is not shown; but one is generally situated somewhere between the smokebox and the injector. The water regulation is done entirely by the handle K, and not by the ordinary water stop cock.

The injector can be changed from exhaust to live steam whilst working by opening the valve X, and then partially closing the wing valve.

Under ordinary circumstances no running at the overflow will take place; but should there be grit or other obstruction in the injector, or should the temperature of the feed water be too high before it approaches the injector, then running at the overflow may be expected. Also in the supplementary portion, the same thing will occur if the spring on the overflow valve has become weak.

It is necessary that the action of the automatic supplementary steam valve should be prompt and instantaneous, and that when in the lifted position shown there should be no escape of water or steam through the small aperture to the right. To ensure this, the seatings of the little piston are ground to a good fit with emery powder; the outer one bearing slightly harder than the inner ring, so that the bearing shall be about equal when the inner one gets warmed up by the steam. Should grit get between the seatings and render the piston leaky, it may be removed by shutting off the boiler steam and flooding the injector with

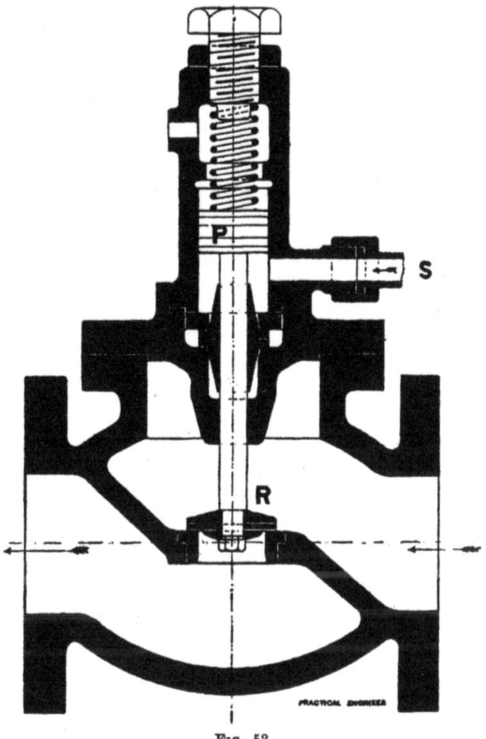


Fig. 53.

water; but should this not prove effective, the injector may be temporarily worked by manipulating the supplementary live-steam supply by hand, by means of the boiler steam valve. When the engine returns to the running shed the obstruction may be removed.

Through the sediment deposited in the cones of injectors working in districts where the water is bad the cones have to be periodically removed, and the scale cleaned off the internal surfaces. Much the same thing happens, though less in degree, when bad oil is used for lubricating the cylinders. To facilitate the operation of removing the scale, Messrs. Holden and Brooke supply cleaning bits of the required shape and size. These consist of a piece of flat mild steel, $\frac{1}{8}$ in. thick, cut in profile to the exact shape of the cone to be cleaned, and in appearance like an ordinary template. The cone should be removed from the injector, and, if necessary, soaked in petroleum, to soften the scale before using the cleaning bits. Care must be exercised not to remove any metal from the cones when using the bits. If required, the cone may be further cleaned by turning a piece of soft wood to the same shape as the bit; but on no account is any grinding material, such as emery powder, to be used with it.

The makers recommend that the water inlet to the exhaust portion should be at least 2 ft. below the lowest level of the feed-water tank; and that the wing valve in the exhaust pipe should be not less than 2 ft. away from the injector for efficient working. When erecting the injector fittings the red lead should be applied to the joints with care, so that dried portions shall not break off and find their way into the cones. An asbestos packing for joints is preferable on this account. After erection or alteration the cones are generally removed, and water blown through the pipes to clear them of all substances that may have got in during erection.

The length of period which an injector will go without cleaning depends entirely upon the quality of the water and steam used. There have been many cases in which years have elapsed without removing the cones. It seems somewhat strange that it is possible for little bits of sediment to stick to the surface of an injector cone when moving at the rate of 100 ft. per second, or more. In the author's opinion, it is not while the injector is at work that deposit accumulates, but at every stoppage a certain amount of water flows through the cones and leaves them covered with a film of water, which, of course, contains sediment in solution. The cones being at a temperature very near the boiling point, this film of water is quickly evaporated, and leaves behind a certain amount of deposit.

In the case of engines working intermittently, such as

winding engines at a colliery, the continual stopping and starting would necessitate the same number of openings and closings of the water stop valve, so that water would not be wasted. This difficulty is surmounted by the application of Johnson's automatic water control, as supplied with Messrs. Holden and Brooke's injectors when fitted to the above class of engines. It is shown in section in fig. 53, and is really a stop valve fitted to the cold water supply pipe. The valve is opened by the pressure of steam on the under side of the small piston P, the steam pipe S being connected to the steam chest of the engine. Directly steam is turned on to work the engine it flows down the pipe S and opens the water valve R, maintaining it open until the steam is shut off from the engine, when the spiral spring maintains the valve on its seating. In this way the exhaust injector is perfectly independent of the attendant, and no water is wasted. These injectors have been fitted to numerous colliery engines, and continue to work well without any trouble.

---

## CHAPTER XVII.

### The Ejector Condenser.

If we were to turn the *whole* of the exhaust steam into an exhaust injector specially constructed for the purpose, we should find that not only would a very large quantity of water be delivered against a very low pressure, but that the condensation of steam would produce a partial vacuum in the exhaust pipe and cylinder. The instrument would then be changed in the character of its object, from that of a forcer of water into a vessel of much higher pressure, to that of passing a large quantity of water through the apparatus against the atmospheric pressure. The principle underlying its action is identically the same as that of the injector; the apparent difference is only one of degree, rather than principle.

With the live steam, or exhaust injector, we mix a comparatively small quantity of water, say 7 lb. or 8 lb., with 1 lb. of steam; and as a consequence, according to the laws of thermo-dynamics, equation (25), the temperature of the mixture is roughly about 170 deg. or 180 deg. Fah. The pressures of vapour at these temperatures are about 8 lb. and 10 lb. per square inch. These are the pressures in the combining cone. By suitable dimensions of the diverging or delivery cone, we may make the pressure in the delivery pipe much greater than either the pressure in the combining cone or that of the source of the steam.

If now we increase the amount of cold water to three or four times the former amount per pound of steam, the temperature of the mixture must necessarily be much less than in the previous case, and consequently the pressure of the vapour must be much less, say 4 lb. per square inch. But on account of the steam having to convey the increased amount of water, the pressure against which it will deliver that water must be considerably less than in the former case.

It so happens that the pressure against which this increased amount of cold water can be delivered is slightly greater than that of the atmosphere; while if we connect the steam orifice of this modified instrument to the exhaust pipe of an engine, the trifling pressure in the combining cone, together with the rapid condensation of the exhaust steam as it comes in contact with the cold water, allows of a fairly good vacuum in the cylinders and exhaust pipe. This, of course, materially adds to the amount of work that can be obtained from an engine; the modified injector performing the functions of condenser and air pump combined.

In this modified apparatus it is necessary to present as much cold water surface to the steam as possible, and at the same time to remove as much resistance to the passage of the water as is convenient in construction, so that the work done by the exhaust steam will be a minimum. To attain this end the steam and cold water orifices change places, and the water flows through the condenser under a small head, or the condenser itself is situated below the surface of the cold-water supply tank.

If we require the amount of condensing water per pound of steam, we merely have to insert the different values for temperatures in the equation

$$\beta = \frac{1114 + \cdot 3\, t_1 - t_3}{t_3 - t_5}$$

assuming the steam to be dry at exhaust. If the dryness fraction is known, then

$$\beta = \frac{h_1 + x_1\, L_1 - h_3}{t_3 - t}$$

When the ejector condenser is supplied with condensing water with a head, then the water cone is so shaped that the flow of water tends to produce a vacuum, in which case the water does not need the assistance of the steam in forcing back the atmosphere; but should the ejector be placed in the water—*e.g.*, in a tank or canal—then the energy of the steam must assist in driving the water through the apparatus. In this latter case the supply of water is greatly in excess of that required for condensing the exhaust steam. Assuming that the ejector receives no assistance from atmospheric pressure, or from a head of water, then, if $v$ = velocity of steam in the steam cone, and $v_3$ = the velocity of the combined jet after mixture, we have—

$$\frac{v}{1+\beta} = v_3.$$

If the combined jet is delivered at atmospheric pressure, through a trumpet-mouth pipe, we must have, neglecting the final velocity—

$$\frac{P_3}{\rho} + \frac{v_3^2}{2g} = \frac{P_4}{\rho};$$

where $P_4$ is the pressure of the atmosphere per square foot. The pressure $p_3$ will range between zero and about 4 lb. per square inch; on the average, say 2 lb. per square inch. If we neglect $P_3$, we shall then have a minimum value of $v_3$, given by—

$$\frac{v_3^2}{2g} = \frac{15 \times 144}{62{\cdot}5};$$

or $v_3$ must not be less than about 47 ft. per second. The maximum quantity of condensing water is then given by—

$$\frac{v}{47} - 1 = \beta \text{ max.}$$

The velocity $v$ may be assumed to be 1,300 ft. per second, and we then find that $\beta$ max. = 27 approximately. When the condenser is fed from a higher level, the quantity $\beta$ may be much greater than 27.

The general arrangement of the ejector condenser invented by Mr. Alexander Morton, of Glasgow, is shown (fig. 54) in end and side elevation. The exhaust pipe E leads directly into the top casting of the ejector T. The condensing water enters by the pipe A, while the condensed steam and water leave the ejector by the pipe C. The main steam pipe from the boiler to the engine is shown at S, and a small supplementary live-steam pipe B, the object of which will be explained further on.

A section of the simple non-lifting condenser is shown, fig. 55. The channel B delivers the water into the throat A, where it comes into contact with the steam, which approaches the water through the channel D. The surface of water exposed to the steam is that of the jet exposed between A A and F F. The delivery cone G is so shaped that the velocity of the water and condensed steam at exit is small. The space surrounding the cone G acts as an air chamber. The rod C C, tapered at its lower end, can be raised or lowered by the hand wheel and screw, and thus the amount of condensing water can be regulated to the amount of steam. The five vanes at R act partly as a guide for the spindle C, and partly to maintain the direction of

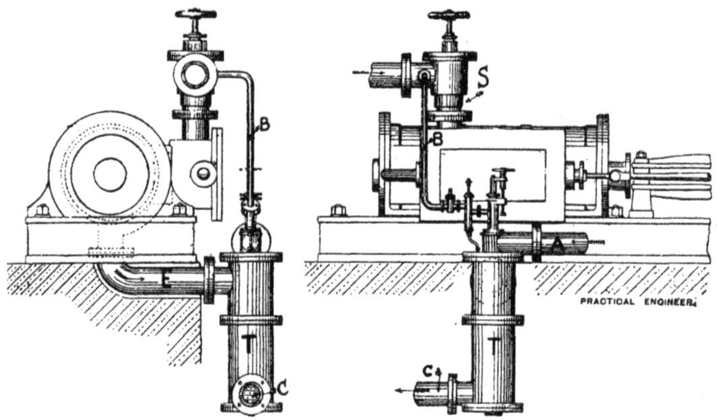

Fig. 54.

the water parallel to the axis of the ejector, so that any tendency towards vortex motion of the stream lines will be checked.

A special form of this ejector is given in fig. 56, in which the water has to be lifted to the exhaust steam cone by means of a small steam jet. The rod C C is made hollow, and perforated near its top end. A small pipe allows steam to flow direct from the main steam pipe into the hollow rod C C, past the two little pistons J, I. The upper side of the piston J is in communication with the atmosphere, and the lower side of I connected to the chamber L K by the small pipe H. If the amount of condensing water is insufficient, there will be a poor vacuum at K, and the pressure in that chamber will assist the spiral spring in forcing up the piston I, so that a free passage of live steam may pass to the ejector. When the steam jet is drawing up too much

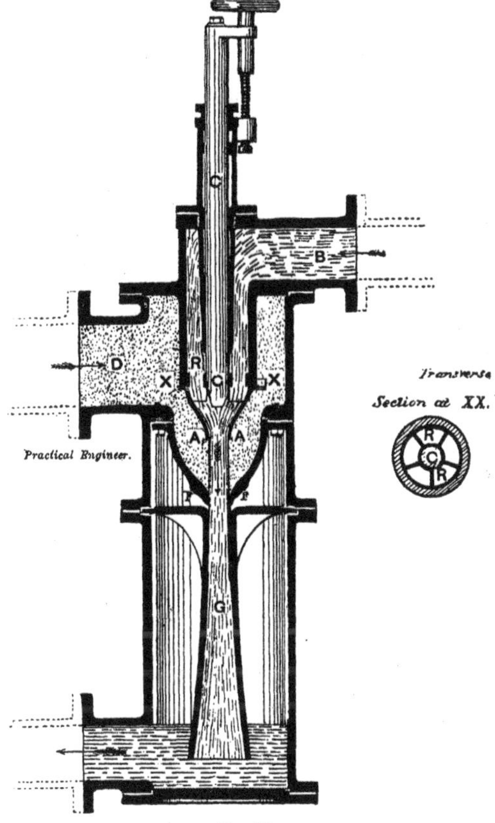


FIG. 55.

water, the vacuum at K will allow the atmosphere to force down the piston J, and close, or partially close, the passage for the live steam. In this way the amount of live steam is automatically regulated to produce the best working

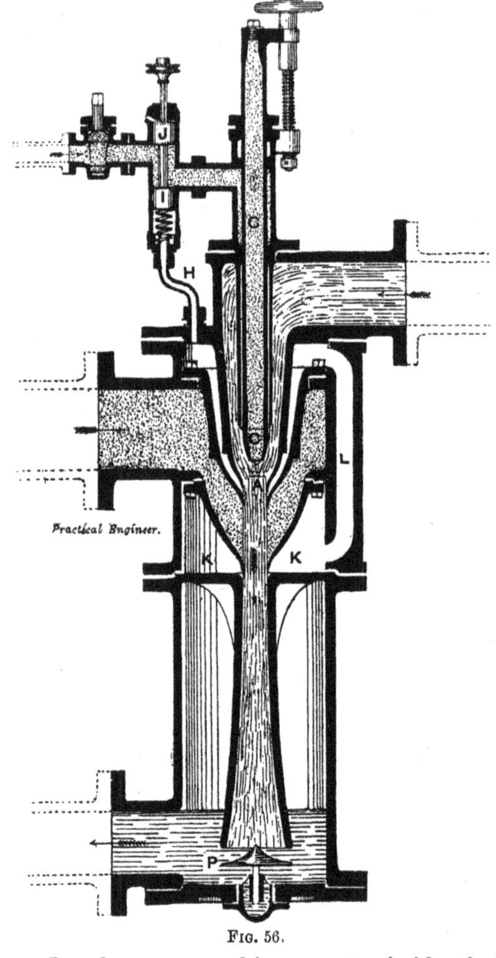


Fig. 56.

vacuum. In other respects this apparatus is identically the same as the one just described.

Another form of ejector condenser is shown in fig. 57, and is manufactured by Messrs. Korting Bros. The exhaust steam enters the condensing cone, and mixes with the water by means of a series of inclined channels drilled in the combining cone, the water having a straight path through the ejector. A non-return valve is generally placed somewhere between the cylinder and the condenser, to prevent any water from being syphoned back into the cylinder.

The makers recommend this fixed capacity ejector in all cases where a head of 15 ft. or more of condensing water is available. The velocity attained by the water is then sufficient to produce a vacuum without the assistance of the exhaust steam, and therefore the condenser is independent of the load on the engine, or the amount of steam used, and will maintain a steady vacuum if it is large enough to take the maximum quantity of steam.

Should the available head of water be less than 15 ft., and the variation of the amount of exhaust steam be small, an

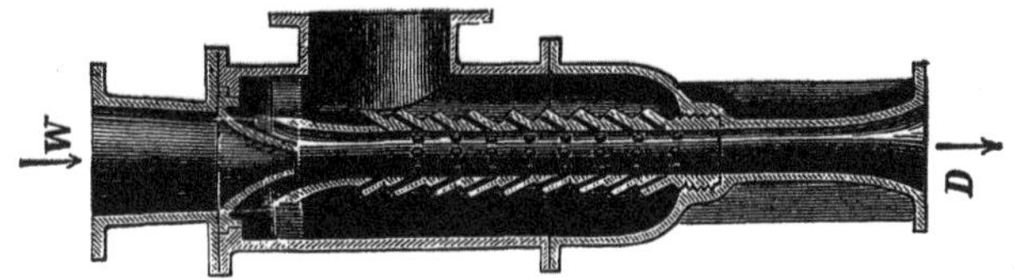


FIG. 57.

injector of slightly different proportions to the one just described may be used, and the injector situated beneath the surface of the water. Such an arrangement is shown in fig. 58, the body of the ejector being at A, the non-return valve at E, the exhaust pipe at G, and the stop valve at F. The flow of water in the river or channel will remove all the hot water and supply fresh cold water to the condenser. Should it be situated in a pond of still water, then the delivery pipe must be made longer, to carry away the hot water.

It may happen that the load upon the engine may vary from time to time, and the supply of condensing water derived from a source which will not allow of the head of 15 ft. previously spoken of. In this case the adjustable capacity condenser should be used, a section and elevation of which are shown in figs. 59 and 60. The construction is something similar to fig. 57, but the delivery tube is made capable of sliding axially, and thus allowing of any desirable number of steam passages being open at the same time. The motion on the tube D, fig. 59, is derived from a connection with the handle C, fig. 60. At D we have a small live-steam pipe for the purpose of charging the

Fig. 58.

condenser when it is situated above the water supply. It may also be used to induce a flow of condensing water while the engine is at work, should it be necessary. The exhaust steam enters at B, while the water comes into the condenser at E. The minimum capacity with this happens when the tube D is moved as far as possible towards E, and *vice versa.*

The variation of capacity ranges between 10 and 100 per cent. These condensers can be fixed either vertically or horizontally, but the vertical position is to be preferred.

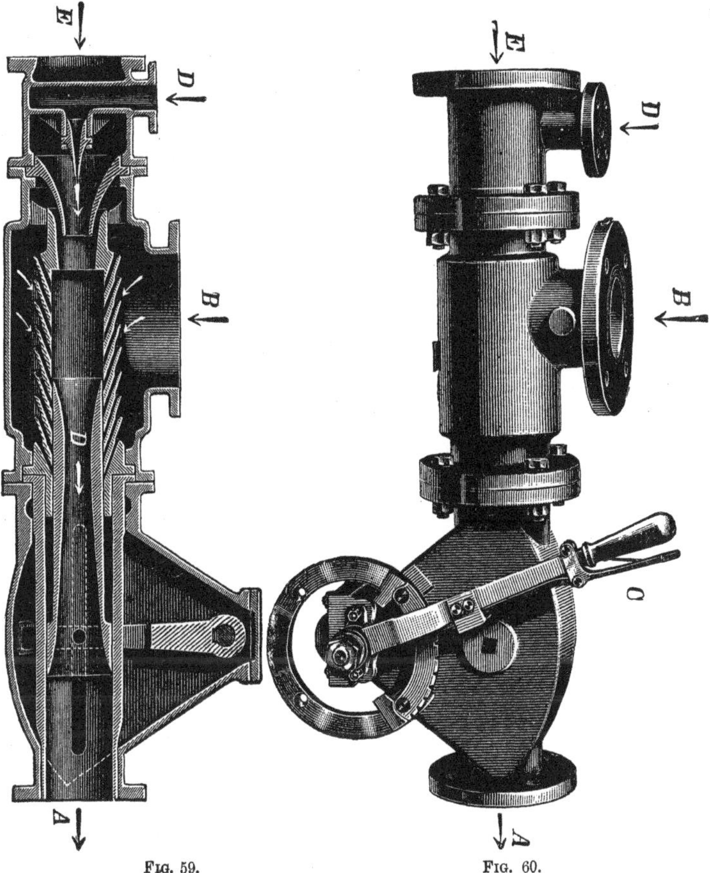


FIG. 59. FIG. 60.

A few results of experiments with an adjustable capacity are given on next page, Table VIII. The water was supplied under favourable conditions, with a head varying from 1 ft.

to 6 ft. The same results also happen with any head under 15 ft. The temperature of the water was 50 deg. Fah.

TABLE VIII.

| Ratio of water to steam. | Position of lever. | Vacuum obtained in inches of mercury. |
|---|---|---|
| 200 | closed. | 23·5 |
| 140 | 1st notch. | 26·5 |
| 125 | 2nd notch. | 27·0 |
| 110 | 3rd notch. | 27·5 |
| 100 | 4th notch. | 28·0 |
| 90 | 5th notch. | 28·0 |
| 80 | 6th notch. | 28·0 |
| 70 | 7th notch. | 28·0 |
| 60 | 8th notch. | 28·0 |
| 50 | 9th notch. | 28·25 |

TABLE IX.

| Ratio of water to steam. | Vacuum in inches of mercury. |
|---|---|
| 160 to 80 | 26·75 |
| 70 to 41 | 25·50 |
| 30 to 25 | 23·50 |
| 20 to 18 | 21·50 |
| 17 to 16 | 20·00 |
| 13 to 12 | 16·00 |
| 11 to 10 | 12·00 |
| 10 to 9 | 8·00 |
| 9 to 8 | 4·00 |

The above figures represent maximum efficiency.

In the above table the position of the lever termed "closed" is that in which the tube D covers up all the small steam channels, and leaves only the top aperture next to the water cone for the inlet of exhaust steam. There are nine sets of small steam channels, corresponding to the nine notches or positions of the lever. In the above experiments the maximum amount of steam was coming from the engine, and hence the best vacuum was obtained with the condenser full open.

It has also been found that when the lever is full open the condenser will work with so little water as 15 lb. per pound of steam, and give a vacuum of from 20 in. to 24 in. Less than that proportion, the vacuum rapidly falls off, and the condenser will soon stop.

The result of some experiments with a fixed capacity condenser are given in Table IX. The head of water was 15 ft., and the temperature of condensing water from 50 deg. to 55 deg. Fah.

---

## CHAPTER XVIII.

### THE WATER INJECTOR.

WITH an ordinary steam injector the motive power is obtained from steam under a given pressure; this pressure being equivalent to a certain head (or height of a column) of steam. If we replaced the column of steam by a column of water we should still have the same phenomenon produced, namely, the motive power fluid (water) taking up water and delivering it against a certain head; but the amount of head against which the water will be forced will be less than when steam was used as the motive fluid, and the amount of water taken up will, of course, be much less.

In fig. 61 the tank A contains the motive power fluid, and the reservoir E holds the water, which has to be pumped up into the tank D. Water descends from the tank A by the pipe B, and is delivered into the trumpet-mouthed tube C, which is in connection with the tank D. Let M N be any datum line, and assume the surface of the water in the tanks A and D to be $z_1$ and $z_4$ respectively above the datum line, while the axis of the injector is $z_3$ above it. Also let the surface of the water in the reservoir be at a distance $z_a$ from the datum line, while $v_1$, $v_2$, $v_3$, $v_4$, and $v_5$ are the velocities of the water—(1) at the surface of the tank A, (2) at the orifice of the pipe B, leading into the injector, (3) at the narrowest part of the pipe C, (4) at the surface of tank D, and (5) at the annular orifice between the pipes B and C.

For the motion of the water between (1) and (2), we must have for steady motion

$$\frac{P_1}{\rho} + \frac{v_1^2}{2g} + z_1 = \frac{P_2}{\rho} + \frac{v_2^2}{2g} + z_2.$$

Now, $v_1$ can be neglected, and $P_1$ is that of the atmosphere $P_a$, the pressure $P_2$ is the same as $P_3$, while $z_2 = z_3$, and hence the above becomes

$$\frac{v_2^2}{2g} = \frac{P_a - P_3}{\rho} - z_3 + z_1 \quad . \quad . \quad . \quad . \quad . \quad . \quad . \quad (74)$$

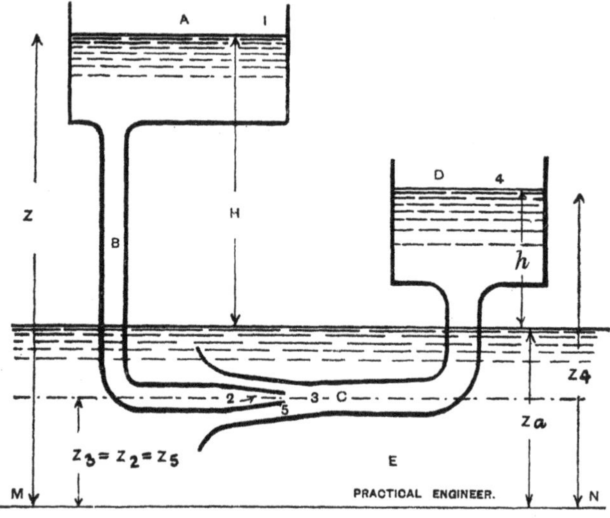


FIG. 61.

Again, between (3) and (4) we must have

$$\frac{P_3}{\rho} + \frac{v_3^2}{2g} + z_3 = \frac{P_4}{\rho} + \frac{v_4^2}{2g} + z_4.$$

We may neglect $v_4$ while $P_4 = P_a$, and thus obtain

$$\frac{v_3^2}{2g} = \frac{P_a - P_3}{\rho} - z_3 + z_4 \quad . \quad . \quad . \quad . \quad . \quad . \quad (75)$$

Also for the bell-mouthed portion of the tube C and the reservoir surface, we have

$$\frac{P_a}{\rho} + \frac{v_a^2}{2g} + z_a = \frac{P_5}{\rho} + \frac{v_5^2}{2g} + z_5,$$

of which $v_a$ is zero, $P_5 = P_3$, and $z_5 = z_3$. After substituting these values, we find that

$$\frac{v_5^2}{2g} = \frac{P_a - P_3}{\rho} - z_3 + z_a \quad . \quad . \quad . \quad . \quad . \quad (76)$$

If we represent the right-hand side of this equation by $x$, while we assume the height of the surfaces (1) and (4) above the reservoir to be H and $h$ respectively, and substitute these quantities in equations (74), (75), and (76), we get

$$\left.\begin{aligned} v_2 &= \sqrt{2g(H + x)} \\ v_3 &= \sqrt{2g(h + x)} \\ v_5 &= \sqrt{2gx} \end{aligned}\right\} \quad . \quad . \quad . \quad . \quad . \quad . \quad . \quad (77)$$

Now, from the laws of impact we have—

$$(1 \times v_2) + \beta \times v_5 = (1 + \beta)\, v_3$$

when 1 lb. of motive-power water from the tank A takes up $\beta$ pounds of water from the reservoir and delivers it into the tank D. Replacing the several velocities by their equivalents in (77), we have—

$$\beta = \frac{\sqrt{H + x} - \sqrt{h + x}}{\sqrt{h + x} - \sqrt{x}} \quad . \quad . \quad . \quad . \quad . \quad (78)$$

As $x$ represents a perfect square, it cannot be negative. With positive values of $x$, we find that $\beta$ increases with an increase of $x$. Writing down the equation—

$$x = \frac{P_a - P_3}{\rho} - z_3 + z_a$$

we see at once that as $P_a$ is constant, $x$ can be increased by either making $P_3$ less, or $z_3$ less, or by increasing $z_a$. The most convenient way of increasing $x$ is the first method, and this may be done by suitably shaping the trumpet-mouthed tube C.

In the above apparatus the work done by the motive power $= H - h$ foot-pounds per pound of water coming out of the tank A; at the same time this pound of water takes up $\beta$ lb. of water from the reservoir and raises it through $h$ ft. into the tank D, doing $\beta\, h$ foot-pounds of useful work. The efficiency of the apparatus must then be

$$\frac{\beta \mathrm{h}}{\mathrm{H} - h} = \frac{h}{\mathrm{H} - h} \cdot \frac{\sqrt{\mathrm{H} + x} - \sqrt{h + x}}{\sqrt{h + x} - \sqrt{x}}$$

which can be written as—

$$\frac{(\sqrt{h + x} - \sqrt{x})(\sqrt{h + x} + \sqrt{x})}{(\sqrt{\mathrm{H} + x} - \sqrt{h + x})(\sqrt{\mathrm{H} + x} + \sqrt{h + x})} \times \frac{\sqrt{\mathrm{H} + x} - \sqrt{h + x}}{\sqrt{h + x} - \sqrt{x}.}$$

$$= \frac{\sqrt{h + x} + \sqrt{x}}{\sqrt{\mathrm{H} + x} + \sqrt{h + x}.} \quad \ldots\ldots\ldots \quad (79)$$

This quantity also increases as $x$ increases. As an example, take H = 150 ft. and $h$ = 40 ft., then from (78), when $x$ is put equal to the successive values 0, 4, 8, 12, 20, the corresponding values of $\beta$ are ·94, 1·24, 1·34, 1·46, and 1·62. Also from (79), when $x$ equals 0, 4, and 8 successively, the efficiencies are ·34, ·45, and ·505, respectively.

The diameters of the orifices may be easily obtained, as we know the several velocities. For instance, let it be required to raise 1,000 gallons of water per hour when H = 150, $h$ = 40, and $x$ = 8. Using the same subscripts as in the previous argument, we have the water raised per second

$$= \frac{1000 \times 10}{3600} \text{ pounds}$$

for every pound of water used from the tank A, there are $\beta$ lb. pumped from the reservoir, and hence $1 + \beta$ lb. that pass through the pipe C. Therefore, the weight of water passing through C per pound of water pumped will be

$$\frac{1 + \beta}{\beta} \text{ pounds.}$$

and weight flowing through C per second

$$= \frac{1000 \times 10}{3600} \times \frac{1 + \beta}{\beta} \text{ lb.}$$

$$= \frac{100}{36} \times \frac{2{\cdot}34}{1{\cdot}34} \text{ lb.}$$

the value of $\beta$ being previously found from above. The volume of the above in cubic feet must be

$$\frac{100 \times 2{\cdot}34}{36 \times 1{\cdot}34} \times \frac{1}{62{\cdot}5}$$

and the diameter of the orifice (3) is given by

$$\frac{100 \times 2{\cdot}34}{36 \times 1{\cdot}34 \times 62{\cdot}5} = \frac{\pi}{4} d_3{}^2 \times v_3$$

or, $$d_3 = {\cdot}042 \text{ ft.} = {\cdot}51 \text{ in.}$$

Again the diameter of the orifice (2) may be obtained thus—

$$\frac{1000 \times 10}{3602} \times \frac{1}{\beta} \times \frac{1}{62{\cdot}5} = \frac{\pi}{4} d_2{}^2 \times v_2$$

hence $$d_2 = {\cdot}0203 \text{ ft.} = {\cdot}245 \text{ in.}$$

The annular orifice (5) must have an area given by

$$\frac{1000 \times 10}{3600} \times \frac{1}{62{\cdot}5} = A_5 \times v_5$$

or, $$A_5 = {\cdot}0057 \text{ square feet} = {\cdot}735 \text{ square inches.}$$

These water injectors are not often used except in isolated cases where a great head of water is at hand. When dock gates have to be examined or removed, the water that has leaked through into the bulkhead compartments must be pumped out. This may be easily accomplished by attaching a flexible pipe to the hydraulic mains and sinking the water injecter below the surface of the water to be pumped out. Of course the injector may be situated above the surface of the water if necessary.

The injectors used for this purpose at the docks at Barry are supplied with water under a pressure of 700 lb. per square inch, and deliver about 5,000 gallons per hour. The maximum lift of the injector is about 10 ft., and the average height above the injector through which the water is forced is 35 ft. These injectors answer their purpose well and require little attention.

# CHAPTER XIX.

## Air Injectors.

The same reasoning may be applied to the pumping of air or other gases by the action of the steam jet.

Let $\beta$ lb. of air be pumped by 1 lb. of steam, the velocity of the steam being V, that of the mixture being $v$, and that of the air as it approaches the steam jet being $V_o$, then we have the equation of momentum—

$$(1 \times V) + \beta V_o = (1 + \beta) v,$$

or,

$$v = \frac{V + \beta V_o}{1 + \beta}.$$

The equivalent head of steam is $\frac{V^2}{2g}$, that of the delivered

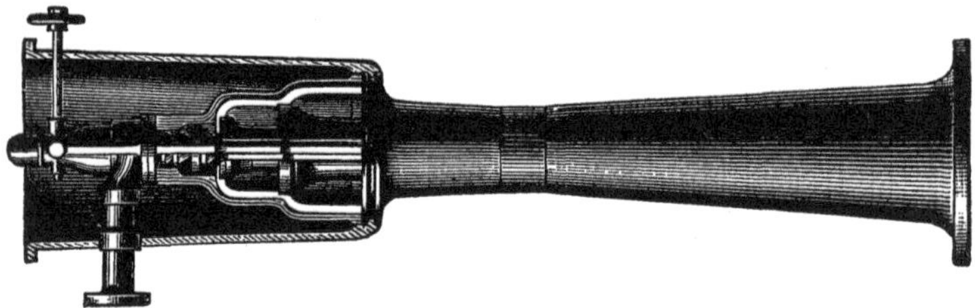

Fig. 62.

air $\frac{v^2}{2g}$. The efficiency of the apparatus as a pump must then be

$$\frac{\beta \frac{v^2}{2g}}{\frac{V^2}{2g}} = \frac{\beta (1 + \frac{V_o}{V} \beta)^2}{(1 + \beta)^2} \quad . \quad . \quad . \quad . \quad . \quad . \quad (80)$$

This shows that the efficiency increases with the velocity $V_o$ of the incoming air. This, of course, will increase with the diminution of the pressure in front of the steam jet orifice, or, in other words, the increase of the vacuum there.

These steam-jet air pumps may be used for two distinct purposes, namely, (1) for forcing or inducing a current of air through chambers or pipes, such as for ventilating or producing a blast, or (2) they may be set to take the air out of a given vessel, and so produce a partial vacuum there. The construction of an example of one of the first of these is shown in fig. 62, the top portion being in section. The double conical trunk contains a series of cones, the first of which is supplied with steam from the small pipe on the left. The small hand wheel is merely for regulating the supply of steam. The use of a number of cones will be readily understood from what immediately follows.

With an ordinary steam-water injector the delivery jet moves down the diverging cone with considerable velocity, generally of more than 100 ft. per second. We could use this jet of water as the motive power jet of a water injector,

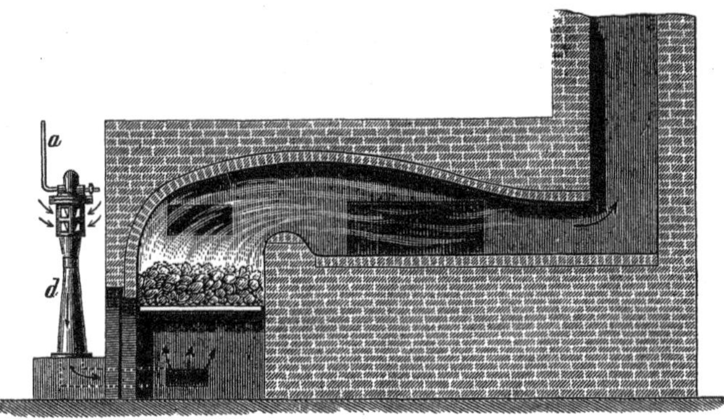


FIG. 63.

and thus pump an additional quantity of water, but, of course, against a less head than when the extra cones were not used.

The head against which air is forced in blowing a fire is at the most only a few inches of water, and sometimes only a fraction of an inch, and it is conducive to economy to force as much air as possible with the least amount of steam, and hence the use of a number of cones to augment the supply of air. The general arrangement of one of these steam-jet blowers, as applied to a heating furnace, is shown, fig. 63, *a* being the small steam pipe; the direction of the arrows denote the course of the air current. The illustrations have been kindly supplied by the makers, Messrs.

Korting Brothers. These blowers are used as a means to produce forced draught, and will, with a pressure of steam of 45 lb., produce a blast of about 2 in. of water. Forced draught enables more fuel to be burnt in a given time than with the ordinary draught, while by its aid the worst of coal and combustible material may be used in the fire.

The amount of air delivered per minute, together with the relative sizes of the apparatus, may be gathered from the subjoined Table X.

TABLE X.

| Purpose of apparatus. | Air delivered per minute in cubic feet. | Diameter of steam nozzle in inches. | Minimum diameter of | |
|---|---|---|---|---|
| | | | Steam pipe in inches. | Air conduit in inches. |
| Undergrate blower.... | 130 | $\frac{1}{12}$ | $\frac{1}{2}$ | |
| | 400 | $\frac{1}{7}$ | $\frac{3}{4}$ | |
| | 550 | $\frac{1}{16}$ | 1 | 9 |
| | 800 | $\frac{1}{5}$ | $1\frac{1}{4}$ | 10 |
| | 1,200 | $\frac{1}{4}$ | $1\frac{1}{4}$ | 12 |
| | 1,600 | $\frac{2}{7}$ | $1\frac{1}{2}$ | 13 |
| | 2,400 | $\frac{1}{3}$ | $1\frac{1}{2}$ | 16 |
| | 3,200 | $\frac{3}{8}$ | $1\frac{1}{2}$ | 18 |
| Ventilator........ | 1,000 | $\frac{1}{8}$ | $\frac{3}{4}$ | 14 |
| | 2,000 | $\frac{3}{16}$ | $\frac{3}{4}$ | 21 |
| | 4,000 | $\frac{1}{4}$ | 1 | 30 |
| | 8,000 | $\frac{5}{16}$ | $1\frac{1}{4}$ | 40 |
| | 12,000 | $\frac{3}{8}$ | $1\frac{1}{2}$ | 48 |
| | 20,000 | $\frac{1}{2}$ | $1\frac{3}{4}$ | 60 |

# CHAPTER XX.

## Air Ejectors.

When the steam jet is used for producing a partial vacuum, it is designed to remove as much air as possible out of a single vessel into which there is no leakage, rather than the carrying of a large quantity of air some distance. An example of this may be cited in the ejector used to exhaust the chambers of centrifugal pumps of air, and so charge them with water at starting. Another instance is the ejector used so universally for producing a partial vacuum in the cylinders and reservoirs of railway brakes. It is for this purpose it has received its widest application, and the brake itself would be of little use were it not for this appliance. A vertical section of the ejector used on the engines of the Great Western Railway is given (fig. 64). The main casting is fixed to the back outside firebox plate, the pipe A passing right through the boiler. Another hole, B, in the boiler plate conveys steam to the channel C, and to the back of the valve D, the latter being operated by the handle H. A port in the valve (for a certain position of the handle) directly coincides with the conical aperture E, whose axis is parallel to the axis of the pipe A. When the valve is turned into this position, steam is allowed to flow through the ejector, and thus exhaust the train pipe F and reservoirs of most of the contained air. The flap valve G prevents a return of the atmosphere into the train pipe when the steam is not passing through the ejector. The brake is applied by a movement of the handle H, whereby ports in the plate K are uncovered and a free passage is provided for the atmosphere into the chamber L, and thence, by a side port (the end of which is shown at M), into the train pipe. The vacuum brake on this railway is generally applied to the train only, the engine being fitted with a steam brake, which is applied and released by the handle (H) at the same time that the vacuum brake is undergoing the same operation. This is accomplished by casting another port, which is controlled by the valve D, which admits boiler steam to the steam brake cylinder.

Another example of an ejector for the same purpose is shown in section (fig. 65), and is made by Messrs. Gresham and Craven, for use with their vacuum automatic brake. This ejector exhausts the train pipe and brake apparatus at starting, and also maintains the vacuum during the time the train is running by ejecting any air which may leak into the train pipe. The first of these operations is performed by admitting steam into the cavity A from which it issues

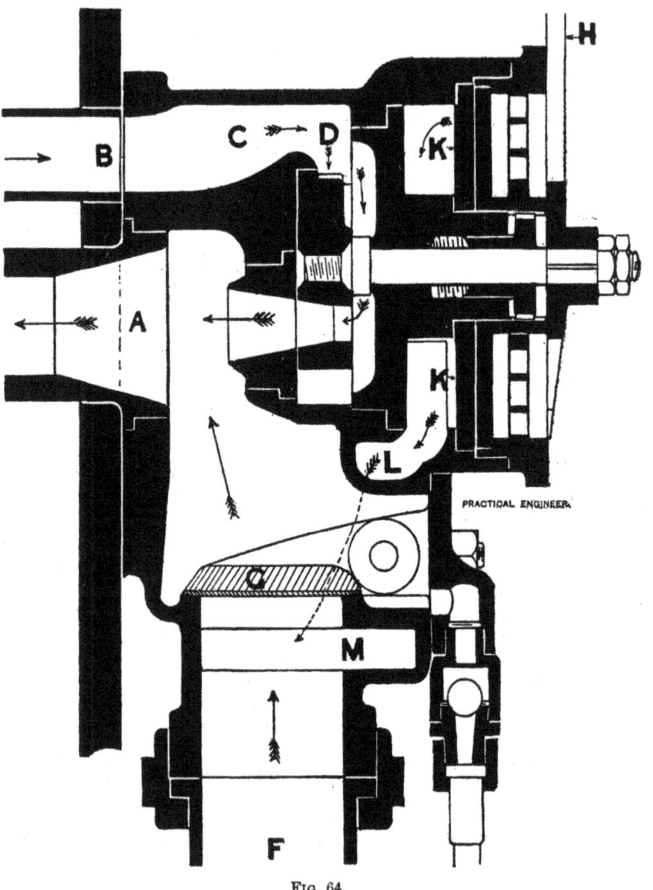


Fig. 64.

into the exhaust pipe B by the small annular orifice C. The diminution of pressure in the exhaust pipe, just beyond

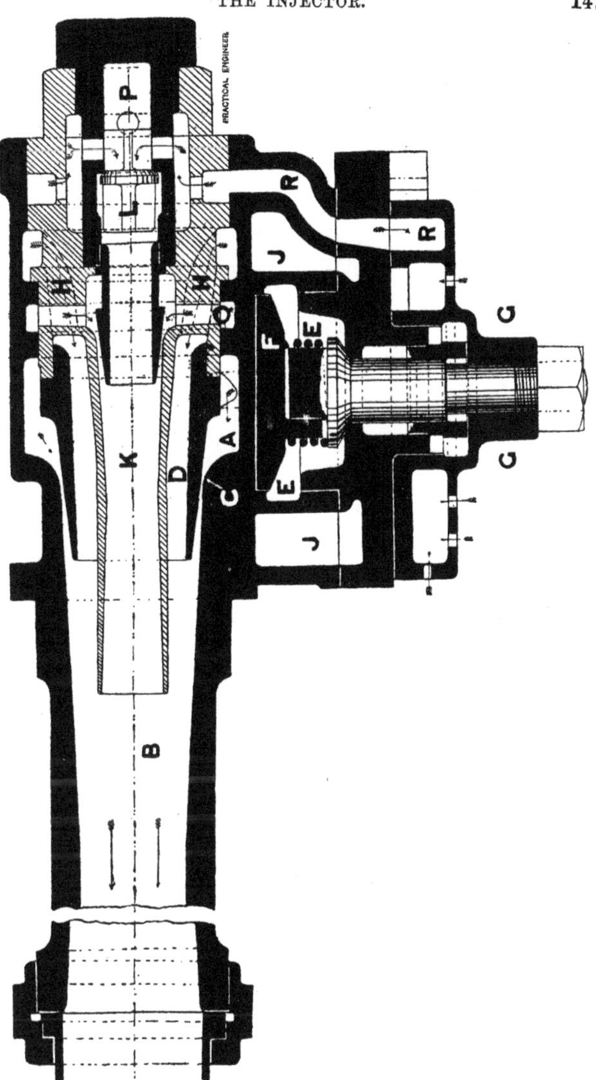


FIG .65.

the orifice, causes a flow of air from the region of high-pressure D to the region of low-pressure B, the air being carried along with the steam into the atmosphere through the pipe B. The space D is in communication with the main train pipe, a non-return valve being interposed to prevent the return of the air when the ejector is not at work. The cavity A is in communication with the steam space E, the supply of steam being controlled by the disc valve F, which is operated by the brake handle. The space E is directly open to the boiler from which it receives its steam.

The brake is wholly manipulated by the brake handle (not shown), which forms part of the casting G, in which there are a multitude of small holes for the inlet of air when the brake is applied. There are three principal positions for the handle, namely, running position, brake on, quick release. In the second of these the handle is turned into such a position as tc admit air to the main train pipe, and at the same time to shut off the supply of boiler steam to the cavity A. The third position necessitates a reversed movement of the handle, while in the first position, named above, provision has to be made for the extraction of any leakage which may take place in the apparatus. This is done by always keeping the small cavity Q in communication with the boiler, so that a jet of steam will continually flow through the annular orifice leading out of Q, and so induce the leakage to flow into the region of low-pressure at K. The valve L prevents the return of any of the exhausted air. The chamber P is in communication with the main train pipe by means of the passage R and the handle disc G.

A side elevation of this ejector is given (fig. 66), with the three positions of the handle G shown. The patches, which are cross-hatched from left to right *upwards*, show the dimensions of the ports in the seating underneath the handle disc ; while the cross-hatching from left to right downwards denotes the ports in the disc itself. To the flange S is attached a small drip pipe.

Fig. 67 shows a transverse vertical section. The steam passage to the large ejector is clearly shown at E, and the channel R leads from the vacuum reservoir on the engine and tender only to the small ejector. The air passage from the disc G to the air pipe is also shown by the arrows. A vertical longitudinal section is given in fig. 68. The valve spindle N regulates the supply of steam to the small ejector.

Release
Quick
G
RUNNING POSITION
N
Brake On
B
G
G
PRACTICAL ENGINEER
S
J
Ports in Seating thus
" " Hand Disc

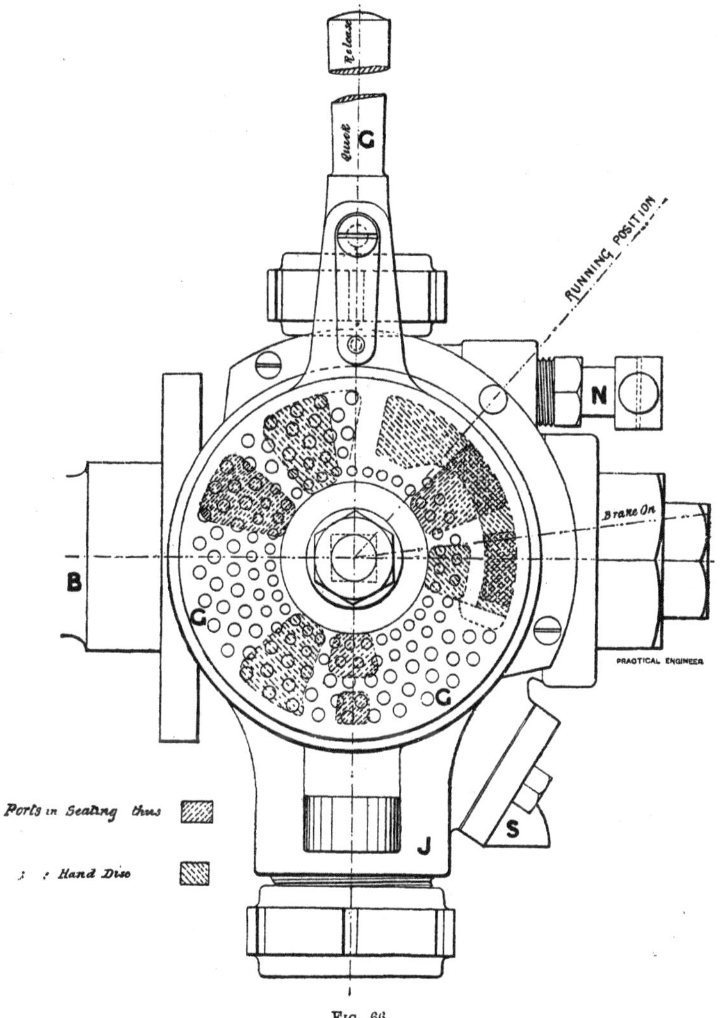

Fig. 66.

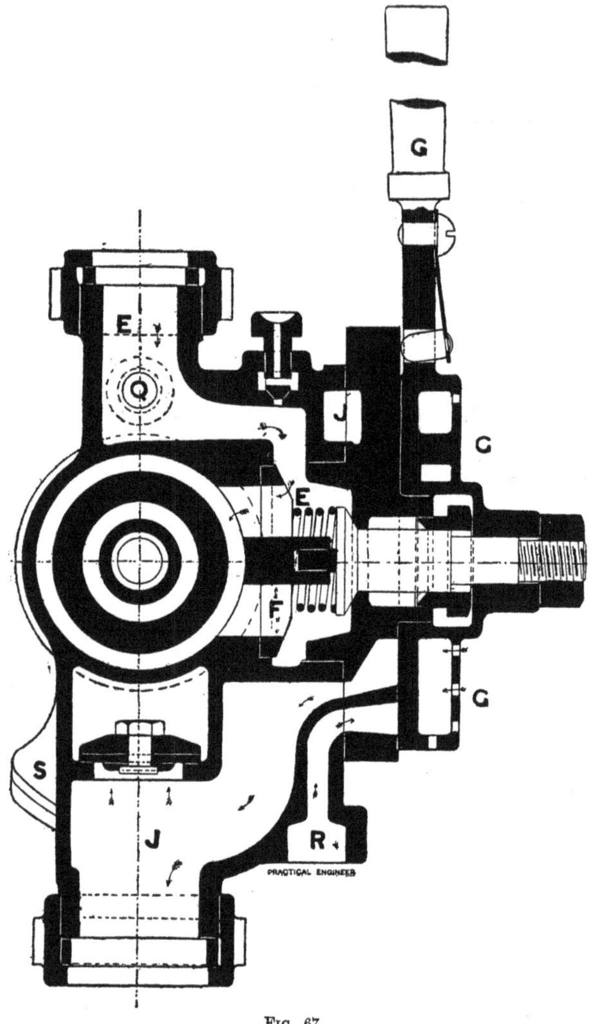


Fig 67.

It is possible with this ejector to produce a vacuum of nearly 28 in. of mercury with a steam pressure of 140 lb. per square inch, but it rarely exceeds 25 in. under ordinary

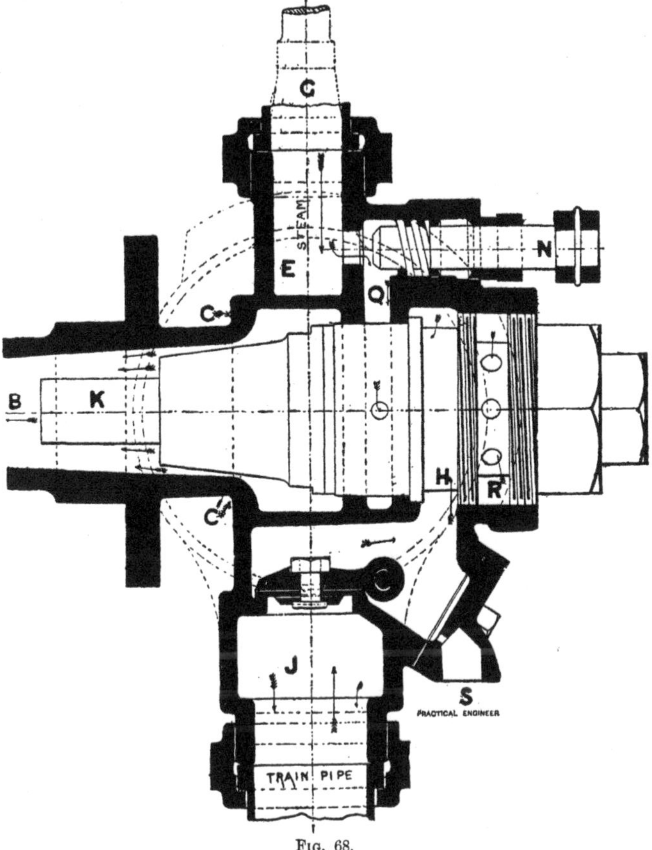


Fig. 68.

working conditions. On more than one railway a valve regulated by a spring is attached to the ejector to prevent the vacuum exceeding 20 in., this being the working pressure in general practice. The reason for this is that, should the ejector be not able to provide so high a vacuum after a particular application that it maintained before the application, the brake blocks would be constantly pressed upon the wheels with a pressure of so many inches of mercury, and the engine would be doing useless work; besides, the longer the train, generally speaking, the more difficult it is to produce a high vacuum. With an engine and eighty-six carriages in a single train, a vacuum of 21 in. has been maintained, the boiler pressure being 140 lb., which seems to indicate that the length of train is practically unlimited as regards the capability of maintaining a working vacuum. A vacuum of 17 in. has also been maintained on a train of fourteen carriages, when the steam pressure was 70 lb., and 18 in. can be obtained on the engine alone with steam at 60 lb. All grease and deposit on the cones have to be removed periodically, and this can be done with an ordinary penknife.

---

## CHAPTER XXI.

### Historical Summary.

(INCLUDING MISCELLANEOUS APPLICATIONS OF THE INJECTOR AND EJECTOR.)

The preceding articles have been more especially devoted to the study of the boiler-feeding injector, and it is principally the development of this apparatus which we now propose to trace out in these few concluding paragraphs.

First made for commercial purposes in 1859 by the inventor, M. Jacques Giffard, in France, it soon found its way into this country, under the championship of Mr. John Robinson, managing director of the firm of Sharp, Stewart, and Co. This firm secured the patent rights for Great Britain, and began to manufacture the injector to patterns obtained from M. Giffard, while at the same time they set about developing new ones from experiments carried out by themselves, the results of which were placed before the Institution of Mechanical Engineers by Mr. Robinson. in January, 1860.

The novelty of the apparatus, and the then apparent mystery which seemed to surround its mode of action, account for the temerity exhibited in the discussion of the paper, in which only such veterans as Sir Frederick

Bramwell and the late Sir William Siemens took part. This may not have been altogether unexpected, if we judge from the concluding paragraph of the paper in question, which runs thus: "It has not been attempted to give any calculations of the power obtained by the injector, and, indeed, the writer has been discouraged from attempting this by the opinion expressed to him by an eminent hydraulic engineer, that the injector is a valuable application of a force which very few persons understand, and which has never been explained in books; and when it was found possible with steam of 24 lb. pressure to inject water into a boiler at 48 lb. pressure, it was felt that it would be premature to bring forward calculations based upon the result of experiments so hastily made, which require much consideration and discussion before any safe conclusions can be arrived at."

From a perusal of the above paper, and of a supplementary paper of a few months later date, we gather that

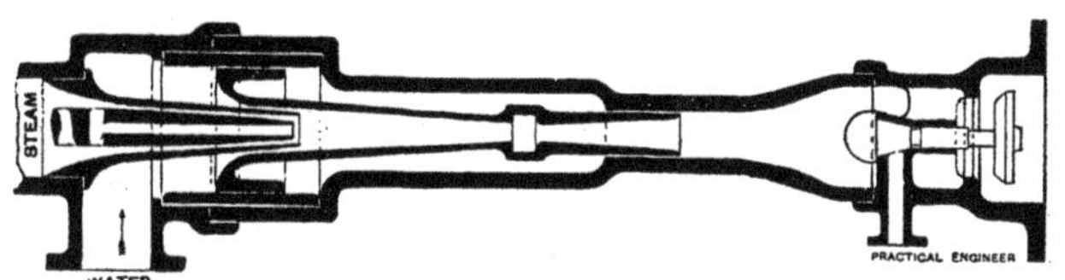


FIG. 69.

the author had not then become acquainted with the laws governing the evaporation of water or condensation of steam; for we find on page 75* that "great surprise was evinced at the remarkable uniformity in the increase of temperature of the feed water in passing through the injector; the average rise being 75 deg., and the extent of the variation being only between 71 deg. and 86 deg., with a range of pressure from 15 lb. to 51 lb. per square inch, and a change of initial temperature of feed from 74 deg. to 110 deg."†

It was left for Sir William Siemens to point out in the discussion for the first time that the theory of the mode of action could be easily investigated by ascertaining whether the quantity of steam condensed in the jet was sufficient to

* Minutes of Proceedings, Mechanical Engineers, 1860.

† These pressures appear to have been measured by an ordinary gauge, and are therefore above the atmosphere.—W. W. F. P.

impart to the jet of water the velocity required for enabling it to overcome the resistance opposed to its entrance into the boiler; for he says: "The velocity imparted would be inversely proportionate to the weights in motion, if there were no loss through friction or eddies—*i.e.*, if 1 lb. of steam, moving at the rate of 1,200 ft. per second, impinged on 11 lb. of water at rest, the result would be (11 + 1), or 12 lb. of mixed water and steam, moving at the rate of

$$\frac{1200 + 0}{11 + 1} = 100 \text{ ft. per second.}$$

This, of course, must follow, because the total momentum before impact equals the total momentum after impact."

Sir Frederick Bramwell supported Sir Wm. Siemens in his remarks, saying that he thought the action entirely due to the velocity imparted to the feed water by the jet of steam, and that it might be illustrated by supposing a cistern of water A with several feet of head to supply a jet in the lower part of the instrument, and another jet from a still higher cistern B to be then brought opposite to the first in the upper part of the instrument, when the greater velocity of the jet from B would necessarily overpower that from A, and force its way into the lower jet. Now, in the case of the injector, if the velocity imparted by the jet of steam to the feed water in the upper jet were greater than that at which the boiler water would escape from the lower tube if unopposed, the water must be forced into the boiler. And since the velocity of the steam was so much greater than that of the water issuing from the same boiler, in consequence of its greatly reduced density, the steam jet was able to impart a sufficient excess of velocity to the feed water to force it into the boiler. Increased velocity was in this case made to produce increased pressure, as in the case of the water ram; only in the injector the propelling steam was all condensed, and got rid of continuously, instead of the actuating water being discharged intermittently, as in the ram."

It is interesting to find, on reference to the above paper, that the proportions of the lifting pattern used at the present time are similar to those used by M. Giffard when he first placed the apparatus on the market, except that we do not now require the little "peephole" directly opposite the overflow gap to satisfy ourselves that the injector is at work. This fact seems to indicate that the inventor carefully perfected his invention before offering it to the engineering public, which precaution no doubt materially promoted its adoption.

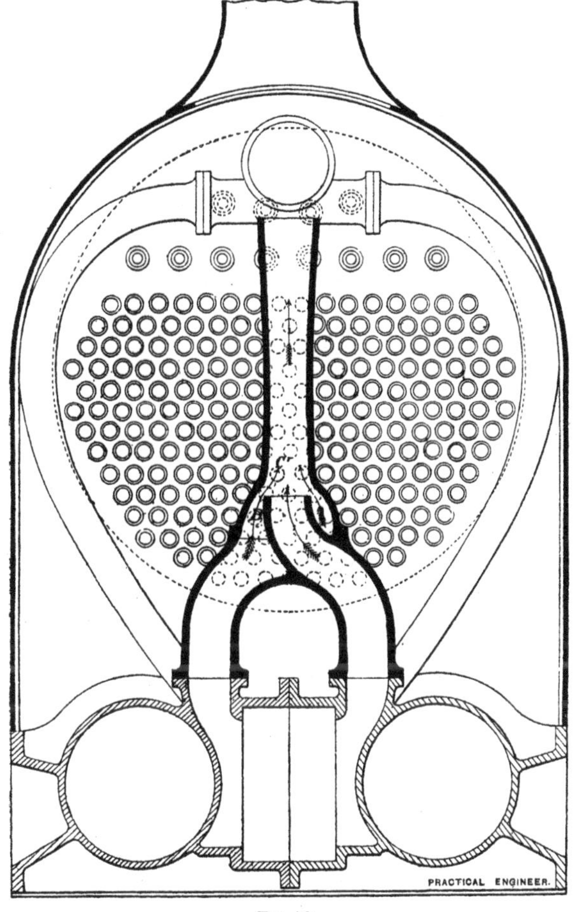


FIG. 70.

The first modification of the original "Giffard" in the

way of simplification seems to have been carried out by Messrs. Gresham and Robinson, in 1864, by making the combining and delivery cones in one casting, and operating it by a rack and pinion arrangement.

About this time Mr. Sellers, of the United States, made an alteration in the original apparatus, the details of which are shown in fig. 69. It is the first attempt that we find to supply automatic regulation. The combining and delivery

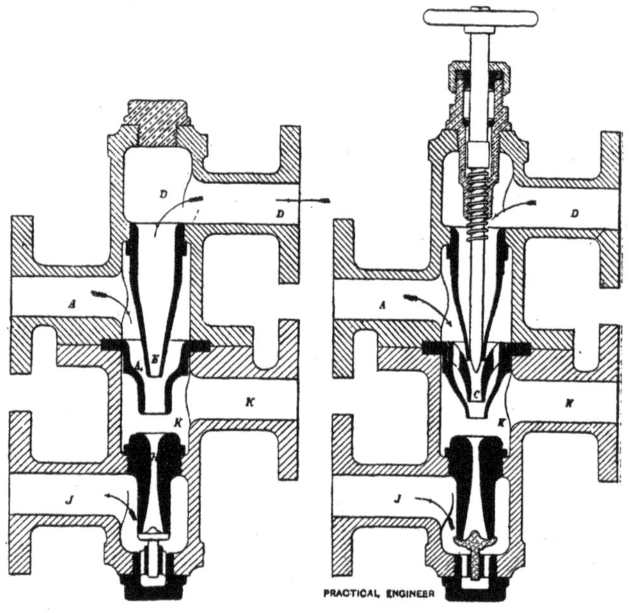


FIG. 71. FIG. 72.

cone, instead of being operated by means of a rack and pinion, have at the feed-water end a piston sliding to an air-tight fit in a metallic cylinder. The usual overflow orifice opens into a chamber, the upper end of which is covered by the above piston. If the supply of feed water is too great, some will escape by the little overflow into the chamber, causing an increase of pressure there, thereby raising the piston and reducing the water supply. Should the water supply be too little, some of the water in the chamber will be induced to flow into the delivery cone through the overflow, and thus reduce the pressure in the chamber, allowing the piston to descend and admit more feed water. The hollow steam spindle first makes its appearance in this injector for the purpose of lifting the feed water at starting; and the first relief valve in the delivery pipe is also contained in this instrument for the

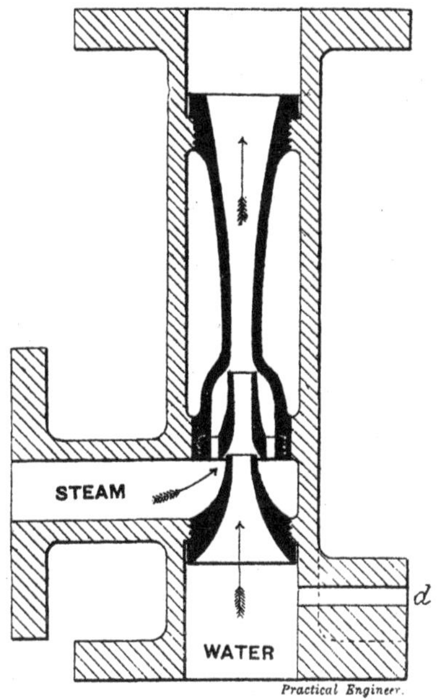


FIG. 73.

same purpose. This injector always delivered the maximum quantity of feed water on account of its automatic water regulation. There does not appear to have been many of these instruments made and used, though the regulation was of the most sensitive nature.

Following soon after this, in 1867, Mr. Alexander Morton, of Glasgow, now senior partner in the firm of Morton and Thomson, obtained letters patent No. 2106, for "Improvements in the Lateral Action or Induction of Fluids, and in the Apparatus employed therefor," which patent is of historical interest, as it includes a description of the exhaust injector and the ejector condenser for the first time.

Mr. Morton begins his specification by claiming that his invention consists mainly in the construction and arrangement of apparatus by which certain currents of either elastic or liquid fluids or of gases may communicate a simultaneous motion to other such fluids or gases by their lateral action or induction through a tube or tubes, increasing in area in a certain ratio to the diminishing pressure of the escaping or acting fluid ; and the objects and advantages to be gained by the use of the said invention are its application to steam engines and injectors, whereby such steam engines and injectors are made more perfect and economical in their working than heretofore.

The first application which seems to have come to his mind was that shown in fig. 70, where the exhaust steam coming from each cylinder of a locomotive was so constrained as to induce some of the steam on the back stroke to flow out of the cylinder, on a principle similar to the modern method of exhausting the train pipes of the vacuum brake, and thus reduce the back pressure on the piston.

We next gather from the specification that in the place of the inducing current of exhaust steam we may, if we choose, substitute one of water at ordinary temperatures, flowing into the apparatus under a small head. In this arrangement, says the patentee, "a full vacuum may be obtained with a very low head of water." In the accompanying sheet of drawings there are two figures (sections) of the above apparatus which are almost identical with the ejector condenser now designed by Messrs. Morton and Thomson, shown in figs. 55 and 56 *ante*. Also immediately following the above we find it stated "that the flow of exhaust steam into the ejector maintains and assists the jet of condensing water as it passes through the apparatus, and that the water jet may be at the first produced by a jet of live steam from the boiler, until the action is fairly started, when the exhaust steam continues the action alone."

We next find a description of an injector by which steam at a very low pressure may be made to force water into a boiler having a very high pressure of steam, in which the actuating steam may have performed a previous purpose before entering the apparatus. Two views of this low-pressure or exhaust injector are shown in figs. 71 and 72, the steam entering by the channel A and the water through the channel D and the central nozzle E. The inventor then goes on to say that "it will be understood, therefore, that in this improved construction of injector the water usually lost at the overflow K is saved, and that by enlarging the area of the annular jet at A, and at the same time diminishing that of the throat H, an apparatus could be made to produce an enormous pressure and velocity of the liquid with a very low pressure of steam."

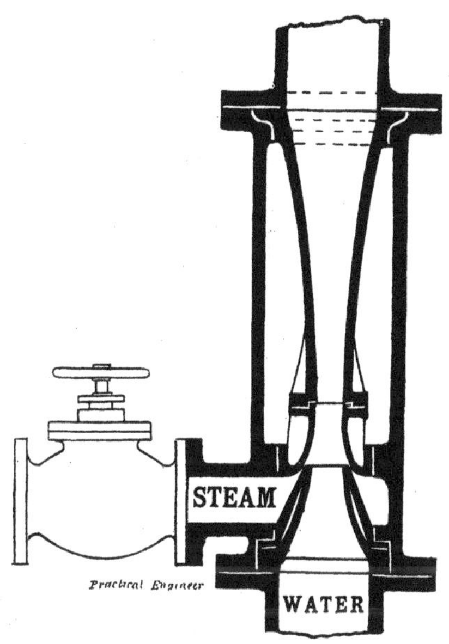


FIG. 74.

The same specification also includes the steam jet pump, fig. 73, commonly called a water lifter, whose principle is identical with that of the injector. For comparison, we may also give (fig. 74) a section of a modern water lifter made by Mr. Morton, and it will be seen that the difference between them is almost inappreciable.

In 1875, a patent was obtained by Mr. Hamer and the brothers Metcalfe for improvements in injectors actuated by exhaust steam, in which both live and exhaust steam were used in an instrument very similar to the original Giffard pattern, a lip being formed in the blast pipe, near the pipe leading to the injector, to catch the steam as it came from the cylinders. The exhaust steam was used after the live steam.

In the following year the same patentees, in conjunction with Mr. Davies, obtained another patent for the use of exhaust steam only, the apparatus being a very close copy of the original Giffard, with an enlarged steam cone.

Mr. Friedmann, of Vienna, whose name has been associated with injectors for many years, was the next to bring out an improvement, which consisted of a double water cone, which separated the water jet into two distinct annular streams,

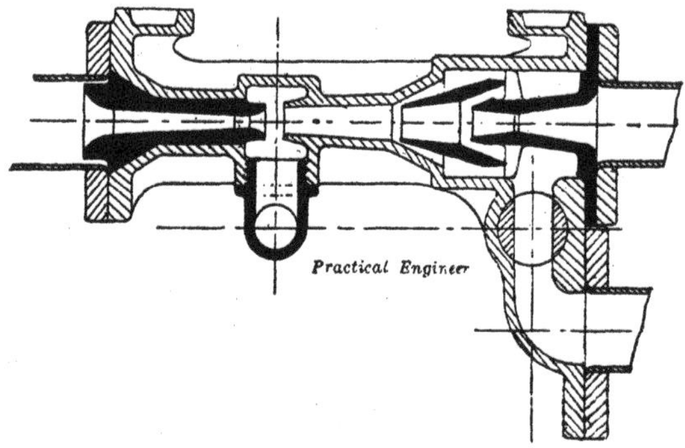


FIG. 75.

for the purpose, it is said, of allowing of the use of warmer feed water than the then injectors were able to take. One of these injectors is shown in fig. 75.

In 1876 the inventor blocked up the waterway to the second water cone, forming a closed chamber *r*, fig. 76, surrounding the first water cone, and drilled a hole O in the combining nozzle, which hole was put into communication with the overflow pipe *v*. By this means, he states that the apparatus is able to take hotter feed water even than before, which circumstance is due to the hole O more than to the air chamber *r*.

In the same specification, No. 4887 of 1877, we find a further and more important improvement, shown in fig. 77, though there are no reasons given for the better result obtained. It consists in putting the chamber *r* in communication with the overflow chamber *i*, and it is obvious that

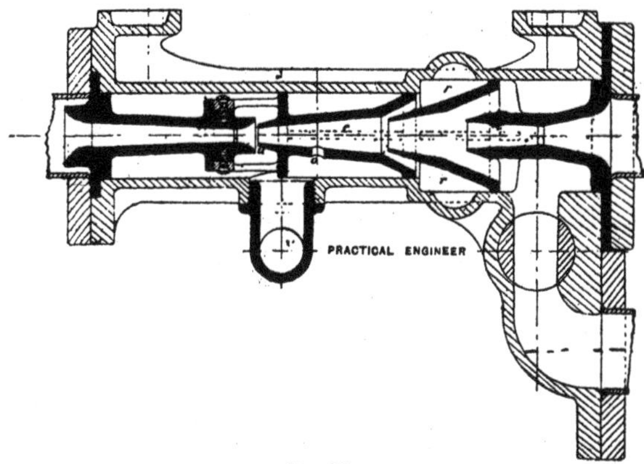


FIG. 76.

it is virtually the same apparatus as that previously described and shown in fig. 19 *ante*, and is the precursor of

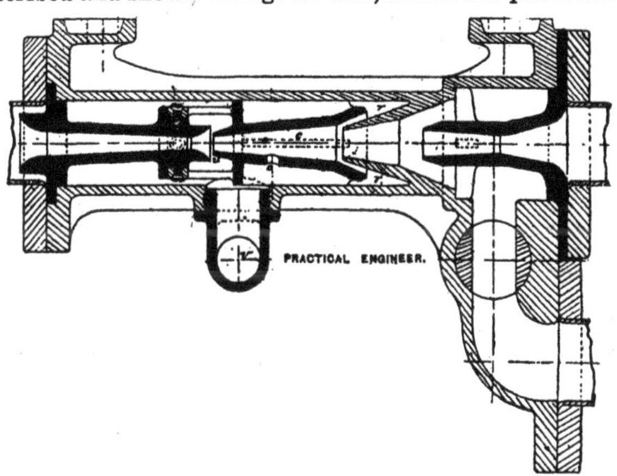


FIG. 77.

all those injectors whose automaticity and re-starting capabilities depend upon a slice being removed from the middle of the combining cone, cut by planes perpendicular to the axis, which gap provides more freedom for the exit of the steam in starting. Of course, these injectors have an overflow valve, otherwise air would enter the combining cone and throw the injector off.

In the same year Messrs. Hamer, Metcalf, and Davies invented the split nozzle exhaust injector, the movable portion of the combining cone being worked by hand, and altogether detached from the main body of the cone, being held tight in its place during work by a series of rods and levers, and was only used at starting. That specification is now out of print, and hence we are not able to give an

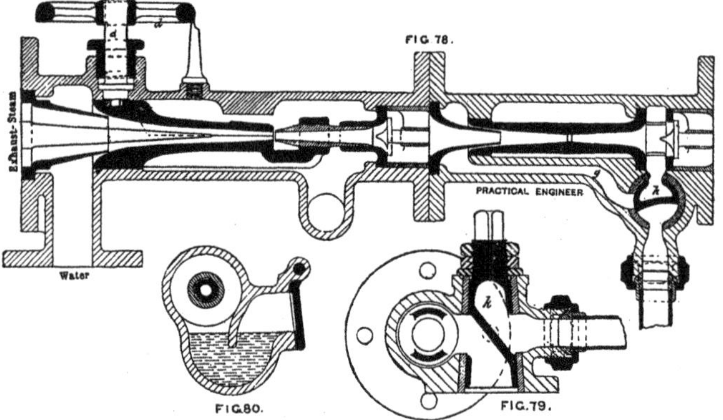


illustration of the rudimentary form of this injector, which has been further developed to so much advantage. There was no outside casing to this injector.

It was not until three years later that Mr. Davies added the hinge to the movable portion of the flap nozzle which is used at the present time. In the same patent we also find the first compound injector recorded by the Patent Office. This is given in fig. 78. Here the exhaust and live steam injectors were placed end to end, the former delivering into the latter, while the live steam for the latter was supplied through the three-way valve K and the channel *g*, entering the injector in an annular jet. The hand valve K, fig. 79, also served as a relief valve in starting. A detail of the exhaust overflow is given in fig. 80. The supply of water is regulated by the hand wheel *d*, which moves the combining cone axially by means of an eccentric pin. The shape of the cones used in this injector has been maintained with hardly any modification up to the present time.

Messrs. Schäffer and Budenberg brought out a modification of the lifting injector very soon after Mr. Davies' compound exhaust apparatus, which had a double water

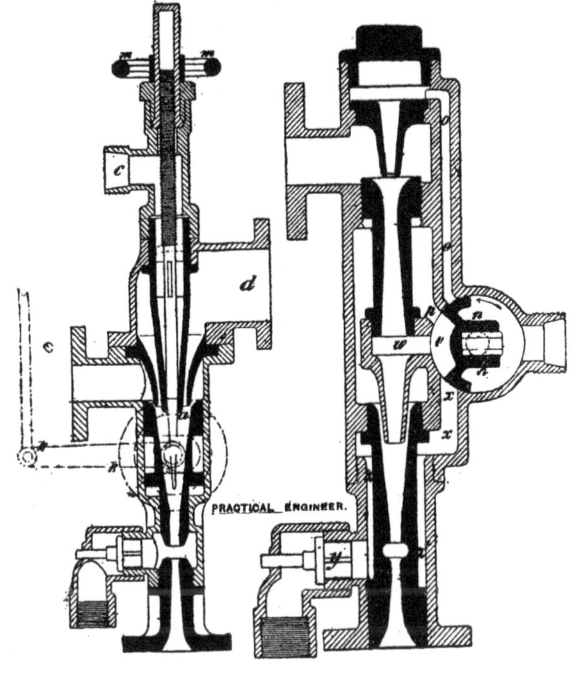


Fig. 81. Fig. 82.

channel and cone very similar in principle to the Friedmann pattern, and for the same purpose. This specification, No. 1593 of 1881, contrary to the general rule, is the essence of brevity, and merely describes the improved portion.

Almost contemporaneously Messrs. Hallum and Smith obtained a patent for the improved compound exhaust injector, shown in fig. 81, and the compound live-steam apparatus, fig. 82. The former of these, the inventors claim, can be used as a simple exhaust, a simple live-steam, or a compound exhaust apparatus. There is also some mention of its simultaneous application as an ejector to aid in producing a better vacuum in the cylinder of an engine. This last statement must be wide of the mark, as the vacuum can only be produced to an appreciable amount by passing the whole of the exhaust steam through the cones; whereas, when water has to be forced into a boiler, not more than about one-seventh of the exhaust steam passes into the injector, and the relative dimensions of the cones for the two operations must be very different. In fig. 81 the combining cone *a* is made to slide axially by means of an eccentric and lever K, which, the inventors say, may be actuated by hand or attached to the governor. This instrument must be very inefficient as a compound exhaust injector, because the exhaust and live steam impinge on the water at about the same place, instead of one after the other, and hence the full amount possible of exhaust steam is not used. The live steam comes in through the pipe C, and the exhaust steam through *d*.

Fig. 82 shows a considerable amount of ingenuity displayed in its design. Steam enters by the three-way valve *n*, thence through the channel *o*, and into the first steam cone. There is no ordinary overflow gap, but a relief aperture is shown at *w*, to be used at starting. The port *p* is also to admit live steam to the body of the injector, and thence by the annular passage into the second combining cone. In the present position of the steam valve no steam can enter either steam cone. To start the injector, the valve is rotated in anti-watchhand direction until the channel *o* is uncovered. Steam then enters the first steam cone, and draws up the water; at the same time the relief aperture *w* is in communication with the passage *x* through the cavity *v* in the valve, and a free passage is provided for the outflowing steam and air to the overflow valve *y*. When the jet has been established, the valve is rotated still further, admitting steam to the port *p*, and closing the relief aperture *w* to the passage *x*.

In 1882 we first hear of Korting Brothers through a patent granted to them for their ejector condenser (previously shown, fig. 57), while they also include an arrangement for regulating the area exposed for condensation by means of an externally-sliding cylinder.

During the following year Mr. R. G. Brooke, of the firm of

Holden and Brooke, took out the first of a series of patents which led up to the present form of their "Influx" automatic injector. The chief improvement lies in the use of the lifting tube A, figs. 83, 84, and 85. In the first of these the steam cone is moved axially by the rack and pinion S and hand-wheel *r*, thus varying the supply of feed water. Of course the chief object to be gained by the use of the lifting tube is to provide a large aperture for the exit of steam and air while exhausting the feed-water pipe, before the injector starts. This aperture is closed after starting by moving the combining cone up to the lifting tube by the hand-wheel *m* in fig. 83, and by the pressure in the delivery pipe in fig. 84. No overflow valve is necessary, as by the closing of the

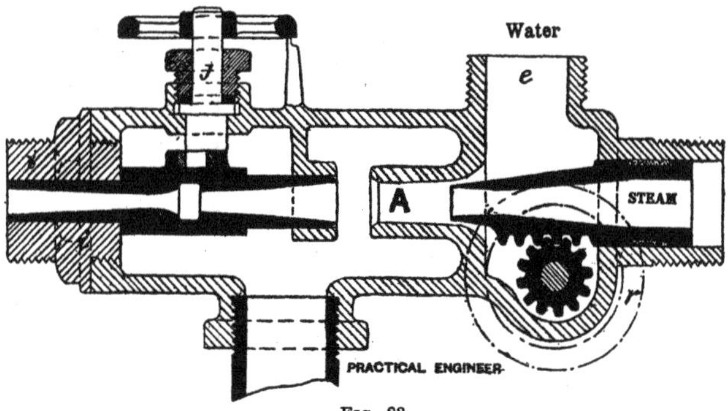


FIG. 83.

aperture no air can get into the combining cone. In fig. 85 all the cones are fixed, and the gap between the lifting tubes and combining cone closed by a couple of spherical valves. It is unnecessary to say that the form of apparatus, of which the last figure is a section, is substantially the same as that manufactured by the firm to-day, for both in the early and modern apparatus there is a gap between the lifting and combining tube which communicates with a small separate chamber, the outlet to which chamber is closed by a valve or valves.

Mr. Gresham, in 1884, introduced his sliding nozzle automatic injector. Two of the original forms are shown in figs. 86 and 87, while the fourth is almost identical with that manufactured by him now, and shown in fig. 26 *ante*. The first of these is very similar to that just described, fig. 84, the spiral spring not being used. In the second form (fig. 87) the delivery cone only moves. On opening the steam valve it forces down the piston head and flows freely from the overflow, but on the arrival of the water the jet produces a great pressure in the delivery chamber, which pressure

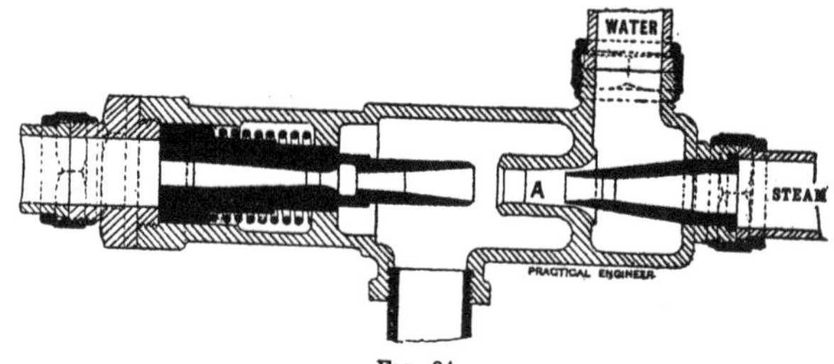


FIG. 84.

forces the delivery cone into its former position, closing the overflow, while the end of the delivery cone comes against the body of the injector and prevents any of the delivered water from returning to the overflow. In this apparatus the cones can be removed without breaking any joints. This system was first introduced by Mr. Gresham as far back as

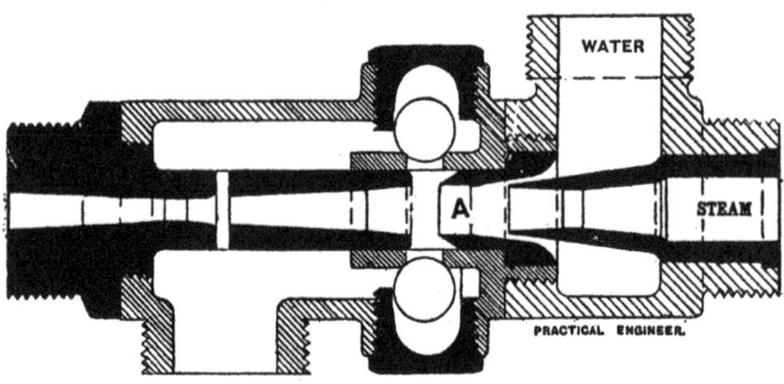


FIG. 85.

1877, and is now a special feature in injectors generally. The same gentleman was the first to combine the injector with its accompanying boiler fittings, and at the same time took the steam for the vacuum brake ejector from the same casting (1885). These are what are now termed combination injectors.

Early in 1887 Messrs. Holden and Brooke adapted their "rigid nozzle" exhaust injector (which they brought out in 1884) to feed against high pressures, such as are found in locomotive practice; and in doing so their chief aim was *to render the live steam supply entirely automatic and independent of the driver or fireman*, so that the change from the previous live steam to the exhaust injector would not entail

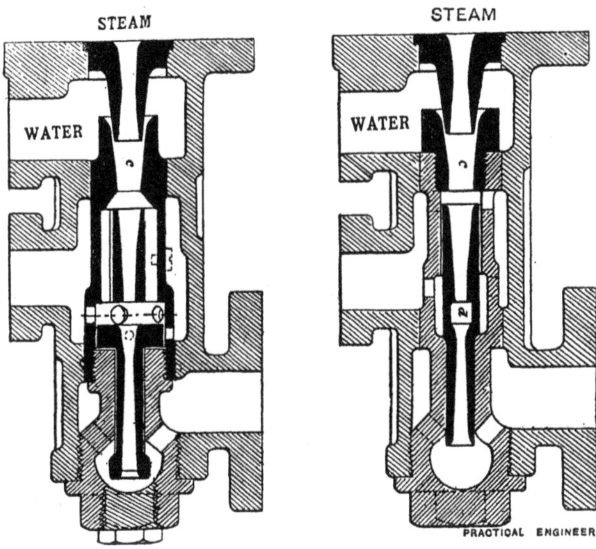


Fig. 87.

Fig. 86.

an increase of movements at starting, or an increase of attention. They have shown a number of different ways in which this can be done, but only a few can be described here. Fig. 88 indicates one method, by which the exhaust combining cone delivers into the supplementary combining cone direct. The supplementary live steam supply, coming from the channel 14, enters in an annular jet, the final delivery taking place through the pipe 5. The overflow to the supplementary portion is closed by a valve maintained against its seating by a spring, on account of the extra pressure and temperature of the jet, which would otherwise cause it to burst. The live steam channel 14 is closed by the valve 12, to the spindle of which is fixed the piston 10. At starting, the exhaust steam and water are turned on, and when the jet has been established in the exhaust portion, and the pipe 5 and its connections become filled, the pressure it produces on the under side of the piston 10 is sufficient to lift the valve 12 against the pressure of boiler steam, and the spring

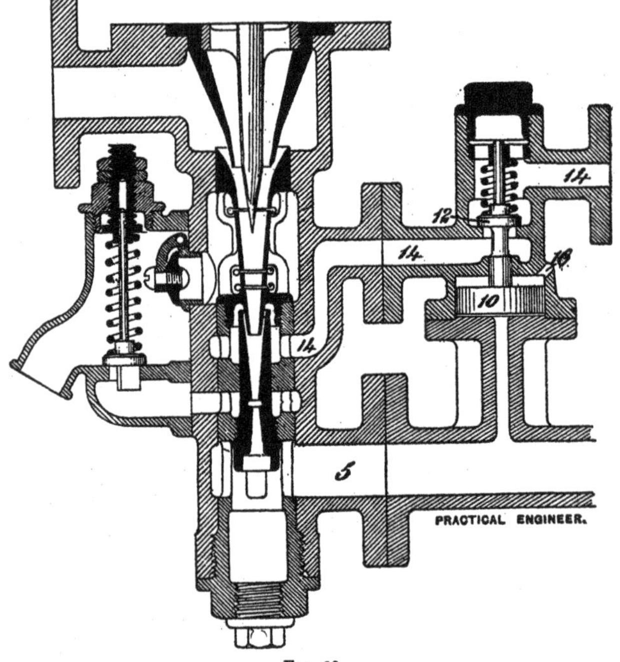


Fig. 88.

thus admitting live steam to supply the additional impulse to drive the water into the boiler.

Another method was to employ a separate injector altogether for the supplementary portion, two of which are shown in figs. 89 and 90. In the former of these the delivery jet from the exhaust portion passed into the supplementary part by the central nozzle, some of which would enter the small port 13, and press upwards upon the flexible diaphragm, thus lifting the live steam valve 14, and allowing the boiler steam to enter the injector by the annular opening 15.

In the other case the exhaust delivery jet lifts the piston 8 with its valve stem, and admits live steam through the first cone ; the water entering, as usual, by an annular opening between two cones.

It is in the same specification that we first find the present compact form of exhaust and supplementary portion in one casting, with their axes parallel. There are also shown other arrangements, something similar to that in fig. 33, with the addition of the automatic steam valve.

The present mechanism for regulating the water and steam by the same handle at one moment is the outcome of

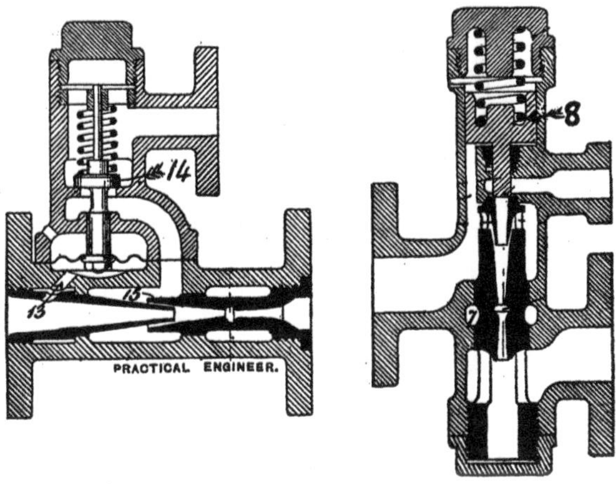


FIG. 89. FIG. 90.

Mr. Brooke's work in 1888, and is now used on all their live-steam injectors.

A series of different relief valves, working in connection with overflow valves, are shown in figs. 91, 92, 93, and 94, all of which are situated in the supplementary injector casting. The object of the relief valve is to assist in establishing the jet at starting or whenever the injector is thrown off, and consists of an opening in the delivery pipe, which is covered by a valve, but the valve is only closed by the application of pressure in the delivery pipe after the establishment of the jet. In fig. 91 the relief valve R is upon the same stem as the piston P, the upper surface of which is in connection with the boiler. The boiler steam will always maintain it open until the pressure in the delivery pipe is sufficient to overcome it and close the relief valve. Fig. 92 is a similar arrangement, but with the piston P cut in two, forming a neck which allows of the overflow being opened to the atmosphere at the same time that the relief valve opens. A spring replaces the steam pressure in fig. 93. A much more complicated arrangement is shown (fig. 94). The

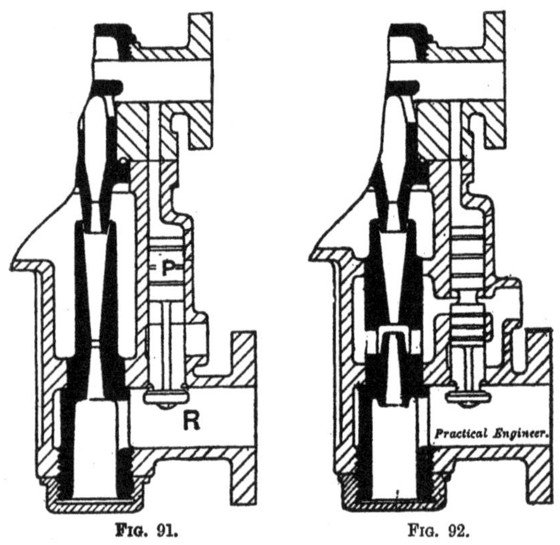


Fig. 91. Fig. 92.

pipes A and B are both open to the boiler, but the former piston is larger than the latter. The two valves are of larger diameter than the pistons, so that when the injector is at work the excess of pressure upon the relief valve C will overcome that in the piston A, and the valve is closed. At the same time the higher pressure on the piston B closes the overflow valve. Should the injector be thrown off, the excess of pressure on the piston A over that on B will lift both valves, and provide a free passage for the flow of steam.

The ejector condenser seems to have thrived only in the hands of a few makers, notably Messrs. Morton and Thomson, the original patentees, and Messrs. Korting Brothers, until the energetic firm of which Mr. Ledward is the head took the matter up. The result was that in 1890 Mr. Thos. Ledward obtained, amongst other things, a patent for steam nozzles of a special construction, and another in 1892 for a variable capacity condenser. The subjects of these two patents are shown in detail in figs. 95 and 96 respectively. In the former of these the water enters at the topmost flange, and passes through a grating formed in the flange of the main converging water cone, and is then delivered by the water cone into a series of co-axial steam cones, which are in communication with the exhaust steam pipe, the water stream and the condensed steam it has picked up being finally discharged through a long, diverging cone. The diverging cone is fixed into the lower portion of the ejector casing, and then the perforated sleeve fitted directly over its upper extremity. The steam cones are then placed in position one by one, their vertical ribs maintaining them co-axial with one another, and also equidistant one from another. The last steam cone being inserted, the converging water cone is then placed in position, thus firmly securing the whole. In all of Messrs. Ledward's condensers are to be found the steam nozzles here shown with vertical ribs. The exhaust steam in all such condensers must approach the

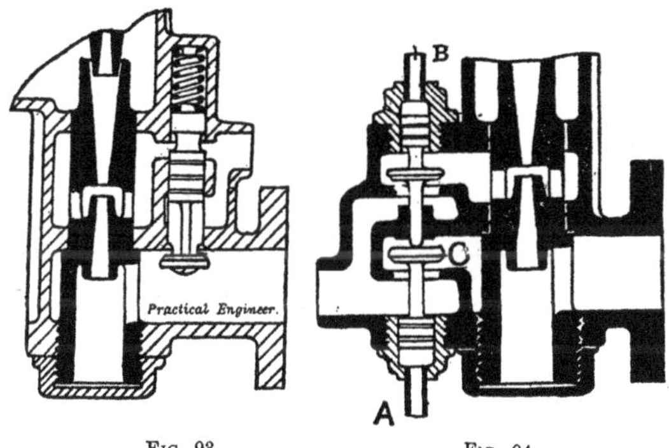


FIG. 93. FIG. 94.

stream of condensing water at a given inclination, according to the shape of the cones. The stream of exhaust steam coming from the engine into the steam chamber of the condenser will necessarily contain eddies, which may distort the water jet in passing through the steam nozzles, or perhaps give the water particles a motion of rotation round the axis of the ejector, as well as a motion of translation along the axis, thus producing in the water jet a sort of screw or tortuous motion. Now, as the total energy in the exhaust steam is only a certain limited amount, if a certain portion of it is used in producing motion of rotation, there

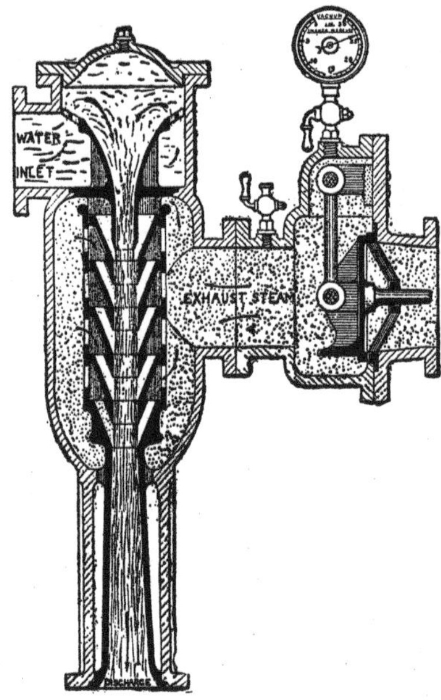


Fig. 95.

remains less energy to produce the linear motion which is required to discharge the jet against the atmospheric pressure, and in this way, therefore, the action may be weakened to such an extent that the condenser may give out. Again, the motion of rotation of itself also tends to break up the jet and produce extra pressure where as little as possible is required. The ribs cast upon the nozzles prevent this distortion of the water jet, and therefore tend to the preservation of its action. The perforated sleeve surrounding the cones is for precisely the same object.

It will be noticed that the steam space between the cones is very large, so that any fragments of packing or sediment which may be carried along with the steam cannot choke up a portion of the space, and thus allow of the free action of the steam on only a portion of the circumference of the water jet, which would tend to distort it. This sort of thing is much more likely to occur when the steam has to pass through sets of small inclined holes, which perform the functions of the cones. This firm has paid great attention to this, and a considerable amount of experiment and experience has strongly influenced them to adopt the particular nozzle now supplied with their condenser.

Now, as was previously mentioned, when the condenser is supplied with cold water from a source whose level or

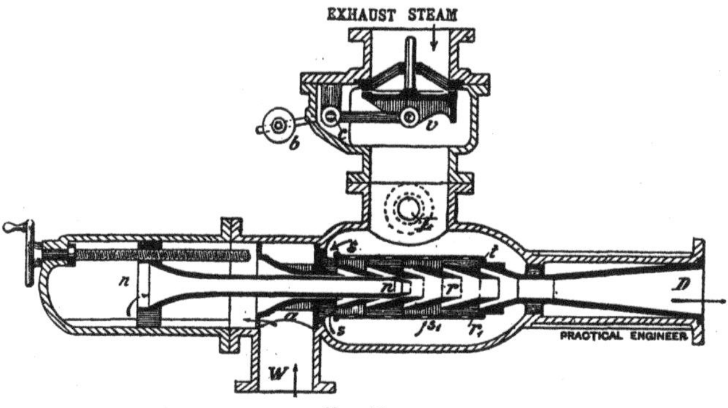


FIG. 96.

equivalent head is not less than about 15 ft., the ejector will continue to work with a very considerable range in the amount of exhaust steam supplied to the condenser, from which we gather that the velocity of the water alone, due to this head, is sufficient to produce a fairly good vacuum without deriving much aid from the exhaust steam—a circumstance which was pointed out mathematically when dealing with ejector condensers. Should the head of condensing water be much less than 15 ft. on the average (Messrs. Ledward have obtained a vacuum of $28\frac{1}{2}$ in. with a head of 12 ft.), it must either be supplied under pressure to the condenser, or the requisite additional velocity must be supplied to the water by the impulse of the exhaust steam itself. The latter method is the one generally adopted.

A nearly perfect vacuum exists in the water jet itself while passing through the steam cones, and hence the supply of water to the condenser will be approximately constant. Suppose that for a time less than the normal supply of exhaust steam comes from the engine. The diminution in the steam with constant water supply, of course, renders the ratio of water to steam much greater, and therefore increases the vacuum. Now, Mr. Ledward has found that as the vacuum is continually increased by diminution of steam, there comes a point above which the vacuum cannot rise, while the exhaust steam maintains a continuous flow of the water jet, when the ejector is depending entirely upon the exhaust steam to keep up the velocity of the jet. This point is with pure steam at a vacuum of 27 in., or a little more with a high barometer. At this point the steam is so attenuated that its impulse is insufficient to keep up the requisite velocity of the jet, and the result is that the atmospheric pressure on the discharge end prevails and forces back the water into the condenser, stopping the action altogether. Now, by reducing the surface of the water jet exposed to the steam, the condensing power of the jet is reduced, and therefore the vacuum will also be reduced, and it is possible in this way to so accommodate the condensing surface to the amount of exhaust steam that the vacuum will be prevented from rising to such a degree as would render the apparatus inactive. If there is much air mixed with the steam, the critical point is lowered from 27 in. to 26 in. In practice it is rarely attempted to get more than a working vacuum of 24 in., and hence a non-adjustable drowned condenser may have its supply of steam reduced about one-half without the apparatus ceasing to work.

To get over the above difficulty when the head of condensing water is small, or nothing at all, the adjustable condenser is used, that one shown in fig. 96 being the invention of Mr. Ledward. The central water cone and tube *n n* is made adjustable by the hand-wheel and screw, so that any portion of the whole number of steam cones may be rendered inoperative by screwing down the tube. It is claimed for this particular apparatus that the inside of the steam orifices are nearer the jet of water than in that of any other design, so that when drawn out to its full capacity the long stream of water receives more support, and is therefore less liable to become distorted than in other systems. Another feature in the condensers, figs. 95 and 96, is the balanced non-return valve *v*, which prevents the cylinder from becoming flooded when the engine stops. The balance weight $b$ just keeps the valve against its seating when not at work. Should the condenser be difficult to start, live steam may be turned into the condenser through the opening L.

Following on with his improvements in exhaust and re-starting injectors, Mr. Brooke, in 1891, adapted the latest

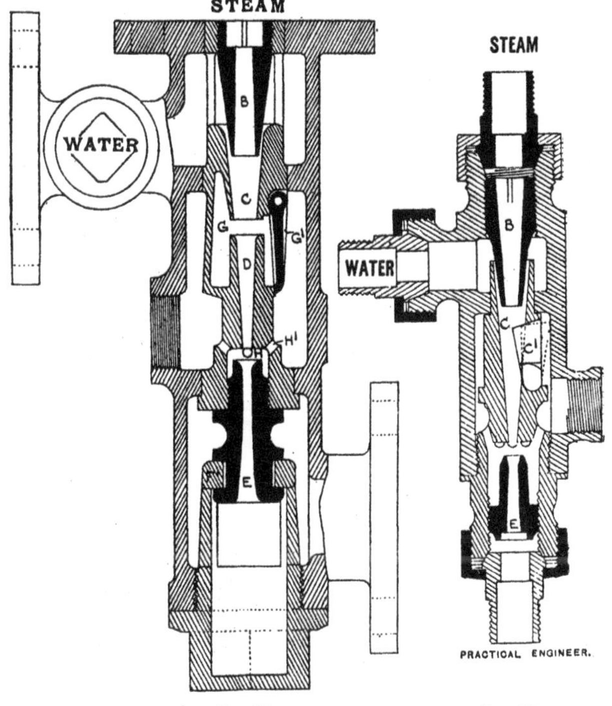


FIG. 97. FIG. 98.

type of his regulating "Influx" injector, to be fitted to the back firebox plate, thus forming the combination pattern shown in figs. 20 to 24 *ante*.

At the same time Mr. Hopkinson, of Nottingham, patented his form of automatic injector, shown in figs. 97, 98, and 99, which in principle is identical with that of Holden and Brooke, and first patented by Mr. Brooke in 1883; but the present improvement consists in casting the walls of the chamber G together with the lifting table C and the combining tube D all in one piece, so that they may be

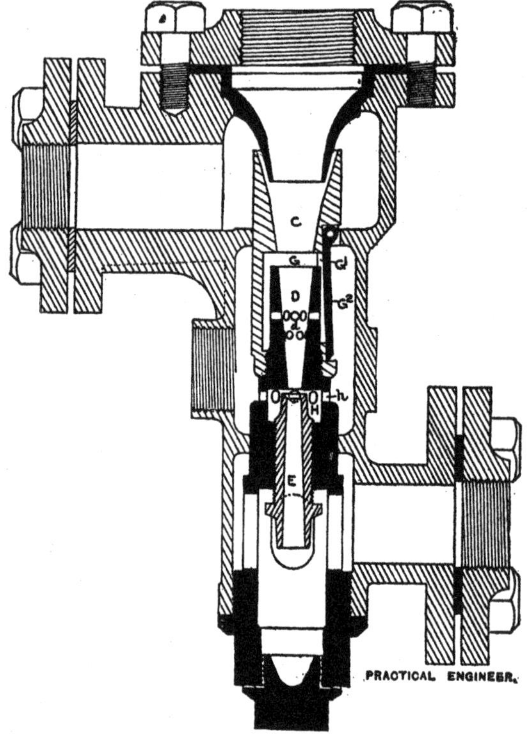


FIG. 99.

withdrawn with the valve plate for inspection or repairs without removing the outer casing. In fig. 98 $C_1$ is a loose portion of the combining cone, the top end of which can fall back against the outer casing, forming a gap in the cone. The piece $C_1$ has its lower end partially cylindrical, and fits into a socket in the cone flange opposite the overflow pipe, which acts the same part as a hinge. This improvement appears to the author to be nearly identical with Davies and Metcalf's flap nozzle.

Mr. Hopkinson's improvement applied to the exhaust injector is shown, fig. 99. Besides the slit G we have two rings of holes $d$ in the combining tube, all covered by the valve plate $G_2$. There are also ordinary overflow holes $h$.

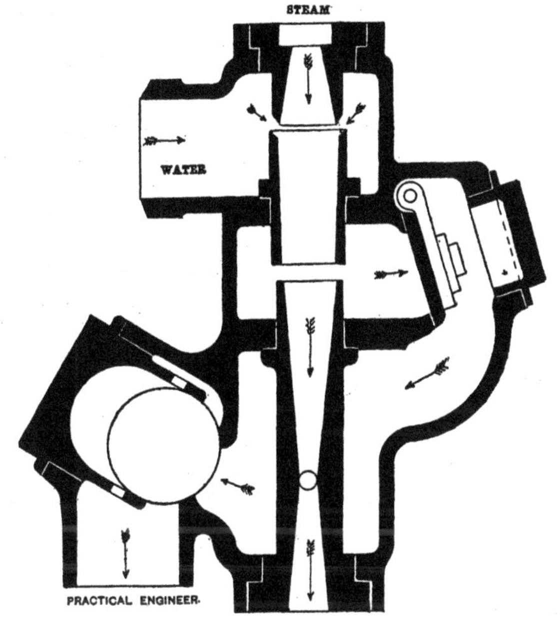


FIG. 100.

Quite recently a patent has been taken out in the United States for an almost identically similar apparatus to Holden and Brooke's Influx injector by Mr. W. Penberthy, a section of which is shown in fig. 100. There are two overflow valves, one covering the aperture of the supplementary overflow, and the other the ordinary overflow. They may be both flap or spherical valves, or one of each, as shown in the figure. The number of the American patent is 484044, of April, 1891.

Another modification of the original Influx injector is shown in fig. 101, and has been patented so recently as December, 1892, by Mr. O. Lindeman, of London. His modification consists of a double steam cone, the outer one

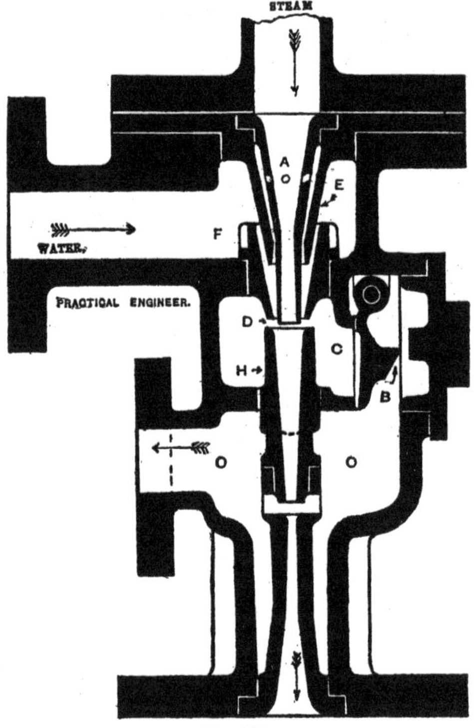


Fig. 101.

receiving its steam from the inner one by means of small holes at A. The inner cone is made to extend just beyond the lower end of the lifting tube, while the outer cone is quite similar to any ordinary steam cone. The combination of parts really forms a compound injector, and if the ordinary overflow pipe were closed by a loaded valve, it should be capable of delivering water at a higher temperature than 212 deg. Fah. The annular jet of steam which proceeds from the outer cone passes through the lifting tube and escapes at the aperture D, and thence past the valve B and out at the overflow O, until the feed water arrives and condenses the steam, when the jet of water delivered by the lifting tube, and already possessing momentum by virtue of the annular steam jet, is further assisted by the steam which issues from the inner nozzle and driven into the boiler. While the annular jet is sucking up the water the jet from

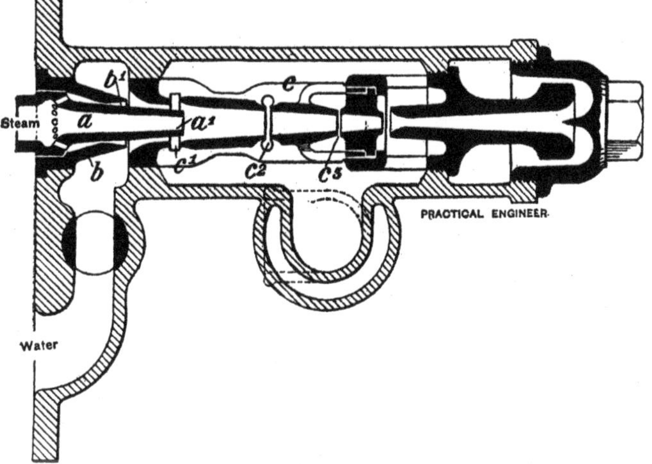


Fig. 102.

the inner nozzle passes out harmlessly through the slit D and the ordinary overflow. The above injector is also, to a certain extent, a self-regulator, the amount of water passing through it depending upon how much is propelled through the lifting tube by the annular steam jet; so that if the pressure of steam falls the amount of water supplied by the annular jet also decreases.

Almost within three months of the sealing of the above patent another was granted to Mr. Friedman for a very similar apparatus, shown in section, fig. 102. The perforations in the inner steam cone point in the opposite direction to those in the apparatus just described. The lifting tube is exceedingly short, and the combining cone contains a couple of transverse slits $c_2$ and $c_3$ in addition to the ordinary overflow. The patentee specifies that the sum of the areas of the openings $c_1$, $c_2$, and $c_3$ must be equal to the sum of the areas of the openings $a_1$ and $b_1$. The author thinks that the injector would start much more freely if

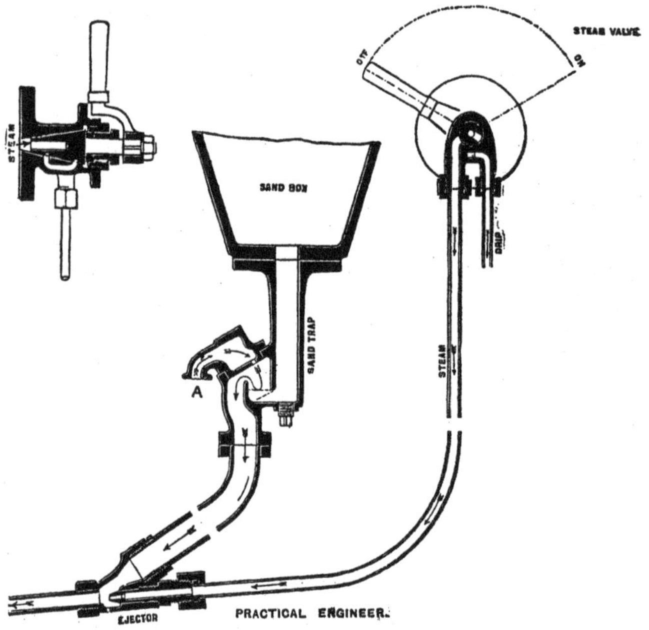


GRESHAM S SANDING APPARATUS.

Fig. 103.

the sum of the former openings were greater than the sum of the latter, the additional area being added to the opening $c_1$. The several portions of the combining cone are cast together, the ribs C maintaining them quite rigid. The overflow pipe has a non-return valve in it which is not shown in the sketch; neither is the delivery flange shown.

Both this specification and the last-mentioned are commendably brief.

Besides the actual feeding of boilers there are many other purposes to which the steam jet may be applied with very considerable advantage, while not only the steam jet, but the compressed air jet also, as applied to the removal of grain from one granary to another or from the hold of a ship to the granary.

The sanding of the rail has been rendered much more complete and efficient since the introduction of the steam jet to the sand pipe by Mr. Gresham, and the sand is no longer distributed over the track instead of on the rail face. A section of the apparatus is given in fig. 103, which also contains the steam valve in transverse and longitudinal

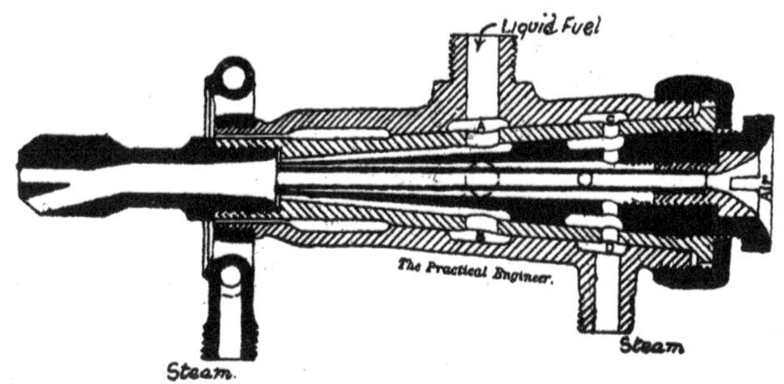


FIG. 104.

section. This valve is so constructed that when not supplying steam to the sanding ejector, any moisture which may leak through the valve is led away down a drip pipe. The jet of steam passing through the ejector induces a current of air down the large pipe joining it to the sand trap, which air enters the hood over the trap entrance by the inlet A, and then follows the arrows in their curvilinear path, at the same time licking up some sand just before descending towards the ejector. When the sand stream arrives at the steam nozzle it is driven forward by the jet down the inclined pipe, which is so directed that the sand will impinge on the rail just in front of the wheel. The inclined iron pipe gets worn away by the sand blast, and has to be frequently renewed.

This apparatus has materially enhanced the value of locomotives with single pairs of driving wheels, allowing of very much increased adhesion at starting and in ascending inclines. The pull on the drawbar, on the average, during a long run is small compared with what it is at starting, and hence the increase of adhesion accomplished by the sand blast makes a simple driver almost equal to a four-wheel coupled engine.

Another extremely useful application of the steam jet is in the apparatus used for burning liquid fuel. The hydro-

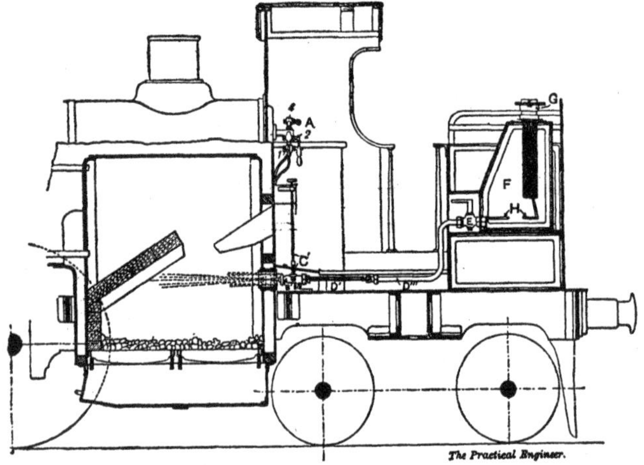


HOLDEN'S OIL-FIRING IN LOCOMOTIVES.

FIG. 105.

carbon refuse and ordinary petroleum require to be broken up into a fine spray and mixed with considerable quantities of air before energetic combustion takes place. The steam jet does both of these operations admirably at the same time by being mixed with the liquid fuel, much the same as it is mixed with water in an injector, and passing the whole through nozzles so as to induce a strong current of air into the combustion chamber. Two methods of accomplishing this are shown in figs. 104 to 109. The first of these is Mr. Holden's system, figs. 104 to 106, as applied to a locomotive and a marine boiler. In addition to the spraying jet, there is a secondary independent annular jet for the purpose of regulating the supply of air to the furnace. The other system is that of Mr. B. H. Thwaite, figs. 107 to 109. He sometimes uses a rifled injector for the purpose of giving the liquid a spiral course, which aids in thoroughly mixing the fuel and air as it is drawn into the furnace. In Mr. Thwaite's arrangement, which is fitted to torpedo boats, and used more especially when full power is demanded, the oil is first sprayed by steam into the hot-air trunk of the forced-draught appliance, and thus it becomes thoroughly intermixed and vaporised before it approaches the fire. Mr.

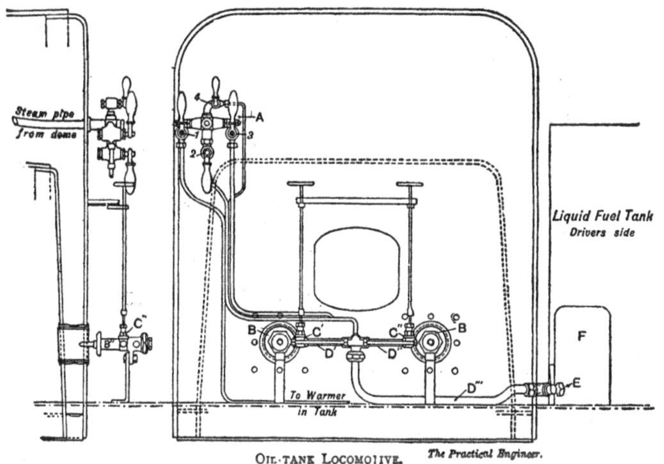


OIL-TANK LOCOMOTIVE.

FIG. 106.

Urquhart has introduced a system of oil-firing for locomotives something similar to Mr. Holden's, and used it with great success on some of the Russian railways.

The efficiency of a Cornish or Lancashire boiler depends in a measure upon the amount of circulation which can be set up and maintained in the water in the boiler. The Acme circulator was designed for this end, and its action is virtually that of the injector. It was described in *The Practical Engineer* of May 19th last.

Ejectors are now used for raising ashes from the stokeholds and throwing them overboard. The apparatus is very similar to the ordinary air ejector.

Professor Carpenter, of Sibley College, Cornell University, U.S.A., has used the injector of small dimensions to determine the amount of moisture held in suspension by steam as it

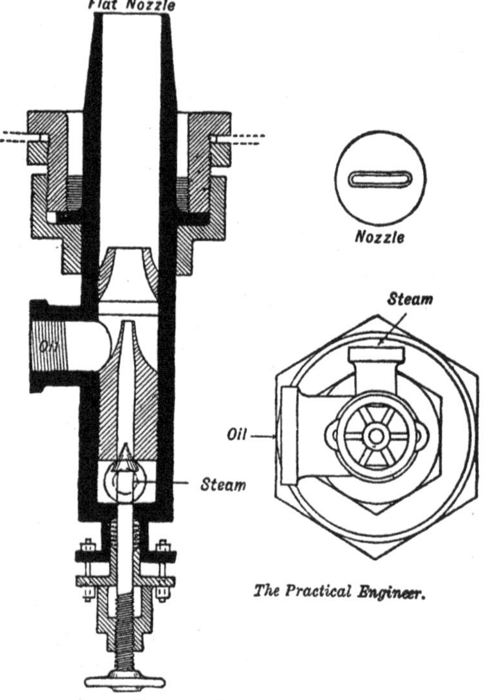


THWAITE'S INJECTOR.

FIG. 107.

passes from the boiler to the cylinder of an engine. A tank of water is weighed, and the injector, which derives its motive power steam from the main steam pipe, draws from and delivers into the same weighed tank of water, so that at the end of the experiment the tank also contains a certain amount of condensed steam. By the use of accurate and delicate thermometers and pressure gauges, and the results of Regnault's classic experiments on the evaporation of water, a determination of the amount of moisture may be obtained.

Other applications of the injector principle may be cited in minute detail, but probably enough has already been given to show its vast utility.

Contrary to general usage, in the previous equations dealing with the live-steam injector, the author has adopted

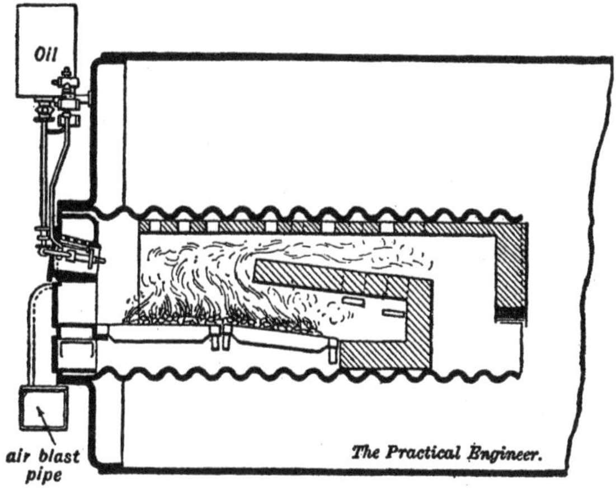


OIL-FIRING IN TORPEDO BOAT.

FIG. 108.

on all occasions an expression for the velocity of efflux of the steam jet, which is that which mathematical reasoning shows should be used, when the difference of pressures causing the flow were such as to produce the maximum weight of steam to flow out per second (generally termed maximum weight-flow), when it is assumed that we are dealing with a so-called perfect gas, the conditions of flow being adiabatic.

Some excuse for this wide departure from the course generally taken needs a little comment, and hence this digression.

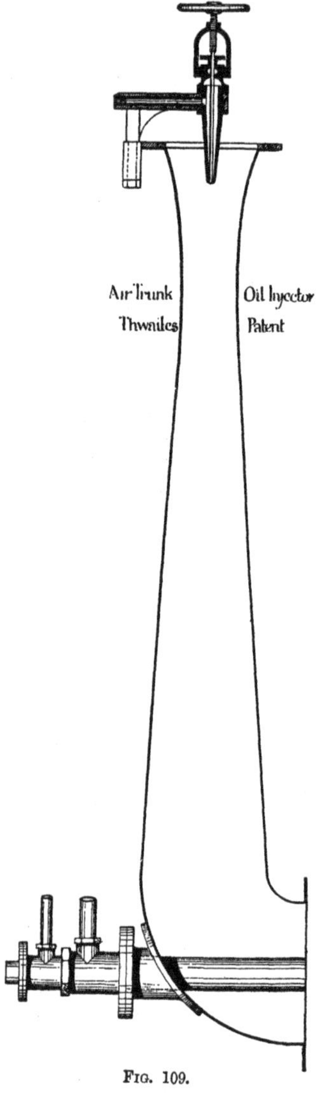


Fig. 109.

It has been previously stated that Mr. Napier found in his experiments on the efflux of steam, that with a constant initial boiler pressure the weight of steam flowing out per second increased as the final or exit pressure was decreased, until this latter was reduced to three-fifths of the boiler pressure, a result predicted by calculation in equation (21). On a further decrease of the final pressure no appreciable change was found in the weight-flow of steam. Hence we may conclude that the maximum weight-flow occurs when the pressure in the chamber receiving the flowing steam is anything between zero and three-fifths of the boiler pressure; or, in other words, under ordinary circumstances the weight of steam flowing out of a boiler per second to an injector will be constant. Now, the impulse of the steam jet is the change in the momentum of its material per second—*i.e.*, the mass-flow per second by its velocity. The former factor we have shown above to be constant, and hence the impulse varies as the velocity. Again, as we have assumed for the sake of simplicity and convenience that we are dealing with a perfect gas, the velocity, according to mathematical reasoning, ought to increase as the difference of pressure increases. But how far are we right in our assumption of the resemblance of steam to a perfect gas? The average steam to be found in a boiler is very far from being a perfect gas, or even a perfect vapour, holding, as it does, a quantity of mist or moisture in suspension. Clearly, then, we must modify the expressions for the velocity of efflux in our calculations to accord as nearly as possible with the actual results, which we can only do by a further reference to experiment. These have yet to be made, but those experiments which have been already carried out upon injectors seem to indicate that the expression used by the author for the velocity of flow is not very wide of the truth, and the error is on the right side.

Should the actual velocity be definitely known in any particular case, it can be substituted for that of maximum weight-flow in the proper equations.

There is another cause for the periodical inspection of steam injectors besides that of deposition of sediment, and it is "pitting." The cones, being made of brass or gun metal, are readily attacked by water at a high temperature containing a small amount of acidity, and in time the once perfect cone becomes a mass of spongy material. Besides the destruction of the metal, we have an exceedingly rough surface, over which the feed water is continually passing at a high velocity; and hence there must be a considerable resistance offered to its motion, which tends to throw the injector off. A thin coating of sediment, of course, prevents this action of the acidulated water, though it also provides a surface which is not so smooth as the pure metal. The cones may be also cut or scored by the passage of grit, derived from impure water, such as would be found in estuaries of rivers. Another cause of stoppage is the local deposition of sediment, which produces eddies; and sometimes bad design will account for local pitting and stoppage.

In illustrating this historical summary, the author has made free use of the drawings which accompany patent specifications, and has selected those which appear to form links in the chain of development of the apparatus.

Imperfect as these notes on the injector are, the author hopes that, in the absence of a work on the subject published in this country, they will still be found of some little use to engineering readers.

John Heywood, Excelsior Printing and Bookbinding Works, Manchester.

# NOTES

# NOTES

# NOTES

www.ingramcontent.com/pod-product-compliance
Ingram Content Group UK Ltd.
Pitfield, Milton Keynes, MK11 3LW, UK
UKHW040556210726
13854UKWH00007B/8

9 780951 936757